CHARLES HOWARD SHINN

LAND LAWS
OF
MINING DISTRICTS

Elibron Classics
www.elibron.com

Elibron Classics series.

ISBN 1-4021-9561-3 (paperback)
ISBN 1-4021-5192-6 (hardcover)

This Elibron Classics Replica Edition is an unabridged facsimile of the edition published in 1884 by Johns Hopkins University, Baltimore.

XII

LAND LAWS

OF

MINING DISTRICTS

"When a new placer or gulch was discovered, the first thing done was to elect officers. In a region five hundred miles long, inhabited by 100,000 people, who had neither locks, bolts, regular laws of Government, military or civil protection, there was as much security to life and property as in any State of the Union. The rights of each digger were definitely marked out and observed."—*Bayard Taylor, in "Eldorado."*

"A special kind of law, a sort of common law of the miners, the offspring of a nation's irrepressible march . . . has sprung up on the Pacific Coast, and presents in the value of a 'Mining Right' a novel and peculiar question for this Court."—*Decision of Chief Justice Chase, December,* 1865. *Case of Sparrow* vs. *Strong.*

"The miners . . . found no laws governing the possession of claims. The reason and justice of the laws they framed challenge the admiration of all who investigate them. Each mining district in an area of not less than 50,000 square miles formed its own rules and adopted its own customs . . . guarding against every form of monopoly and requiring continued work and occupation."—*Speech of Senator Stewart, June,* 1866.

"The bill proposed adopts the rules and regulations of miners." They "are well understood, and form the basis of the present admirable system. Popular sovereignty is here displayed in one of its grandest aspects, and simply invites us not to destroy, but to put upon it the stamp of national power and unquestioned authority."—*Report of Chairman Conners, of the Senate Committee on Mines and Mining, May* 28, 1866.

"Whenever institutions have grown up of themselves . . . they become a matter of scientific study. They are part of the general institutions of the English people, as those are again part of the general institutions of the Teutonic race, and those again part of the general institutions of the whole Aryan family."—*Edward A. Freeman, in his "Introduction to American Institutional History," University Studies, First Series, No.* I.

JOHNS HOPKINS UNIVERSITY STUDIES
IN
HISTORICAL AND POLITICAL SCIENCE

HERBERT B. ADAMS, Editor

History is past Politics and Politics present History—*Freeman*

SECOND SERIES

XII

LAND LAWS
OF
MINING DISTRICTS

By CHARLES HOWARD SHINN, A. B.

BALTIMORE
N. MURRAY, PUBLICATION AGENT, JOHNS HOPKINS UNIVERSITY
DECEMBER, 1884

THE FRIEDENWALD CO., PRINTERS,
BALTIMORE.

LAND LAWS

OF

MINING DISTRICTS.[1]

In the autumn of 1878 the writer, then teaching school in a little Mining Camp of Trinity County, California, witnessed the practical operations of "rules, usages and customs" in regard to local government over well-defined areas known as "Mining Districts," of which there were a number in the county. One, then newly organized, lay partly within the limits of two counties, but it was governed easily and well by the citizens of the District, according to a code whose more important features had been evolved amid the stress and strain of the pioneer days of 1848 and 1849. County officials

[1] This essay forms a portion of an investigation into the entire history of Mining and Mining Camps, ancient, mediæval and modern, with a hope of giving the forms of social organization manifest in the early "Districts" of the Sierras, Coast Range, and Rocky Mountains, their proper place in the story of institutional development upon American soil. Some part of the material here presented will re-appear in a volume entitled "Mining Camps, a Study in American Frontier Government," which will be published early in 1885 by Charles Scribner's Sons, New York. Portions of this work were read last year before the "Historical and Political Science Association" of the Johns Hopkins University. An article upon "California Mining Camps," in the "*Overland Monthly*" for August, 1884, one upon "The Golden Prime of Forty-Nine" (illustrated), which appeared in the "*Magazine of American History*" for November, and one upon "Enactments of the Early Miners," to appear in the *Overland* for the current month, December, comprise the author's publications in this field.

Mr. Shinn's work is a natural, although unconscious, continuation of Mr. Johnson's study of "Rudimentary Society Among Boys," *Studies, Second Series*, XI. This paper might be called Rudimentary Society Among Men.—Ed.

existed, townships and supervisors were parts of the State system, school-districts had been laid out years before, but along Weaver Creek, and in the bends of the yellow and foaming Trinity River, men were still holding "claims" and regulating their business relations with each other by local laws; and in camps further removed from stage roads, county-seats and agricultural communities, the "Laws of the District" covered a broader range, and still dealt with many subjects besides land-tenure. Again, in the spring and summer of 1879, while a special travelling correspondent for the San Francisco Daily *Bulletin*, the writer explored the mining region of the Sierra counties of California, became acquainted with pioneers, and visited many of the famous old camps. Here, also, the influence of the past was, and is, evident in varied and powerful survivals; the institutions that American freemen created in time of need for the protection of life and property, have shown an abiding strength and endurance. This phenomenon of a still surviving and respected local land law deserves more attention than it has hitherto received from historians of the Far West. Viewed in its larger relations, as a curious chapter in the record of social experiments made by men of our Germanic race, the entire code of the Early Mining Districts, or Camps, has a more than local value. In that Code, with its multifarious variations and eccentricities, its curiosities of legislation, its mingling of forms borrowed from alien races, its underlying common-sense and dignity, there is no department more important than that which was devoted to the acquisition and tenure of mineral lands in the several districts at a time when neither State nor General Government had as yet assumed control.

Before entering upon a discussion of the features which distinguish this department of the Miner's Code, we may say, in general terms, that it is difficult to set forth the supreme importance of laws which govern the ownership of land. The social, economic, and political history of the human race has turned upon the pivot of changes in systems of land tenure, and here is a battle-ground of the future, as of the past. Nothing which serves to illustrate the workings of any land system, or of any method by which in practice lands were held, can ever be called irrelevant or worthless, for the entire field of study is so broad, and is broken into so many angles, that each ray of light is needed for its illumination. The best thoughts of writers of the highest ability have been devoted for many years past to studies of early land tenure under simple conditions of life in primitive agricultural communities, and to the potent social and political causes which slowly substituted private ownership of land for the rule of the village and the authority of the old Teutonic field meetings. The study of Germanic land-laws, as taking shape in free discussions of folk-moot assemblies deep in the Swabian Mountains, and as modified in Saxon and Norman England, does undoubtedly appear to offer the most attractive single field of investigation known to institutional history. Nevertheless, we shall find, if we sufficiently examine the local institutions of America, that a contribution of real interest and value has been made to the land-laws of our race, springing into existence naturally, as all such things must, from the character of the people and the exigencies of their situation.

Therefore it is to the American miner of the region west of the Mississippi that we must look for the adoption, in time of need, of this definite and efficient system of local government for hundreds of small communities known as Mining Camps, or Districts, and possessing under a multiplicity of forms those essential safeguards of life and property that the true pioneers, from our Aryan stock, have always been able to secure. To-day, over the Western third of the United States institutional life traces some of its most important elements to the cabin of the placer-miner. Indeed, we may fitly call this the only original contribution of the frontiersmen of America to the art of self-government. Sevier and his Tennesseans did indeed organize their temporary State of Frankland, but it soon and wisely accepted the inevitable. Boone and his brave companions, Kit Carson, Sublette, and his trappers and fellow-guides, Bent, Bridger, Beckwourth, St. Vrain, and their "mountain-men," the companies of voyageurs, the bands of fearless explorers of half a century ago, one and all melted away before the westward-hastening tides of civilization without being forced by imperious necessity to the creation of any code of local laws, or to the organization of any system of permanent legislation. The institutions of the older communities already grown into States had followed too fast in the footsteps of these pioneers. But that army of State-builders who poured out their mighty toil upon the placer mines of the Far West thirty-six years ago, had no sooner pitched their tents beneath the Sierra snow peaks than they called meetings of "all the freemen of the camp," created mining "districts," elected officers, clothed them with sufficient authority, and ordained laws under which peace was secured and prosperity reigned for years.

The purposes to which this "Study" is devoted are narrow and definite. We are necessarily compelled to an entire neglect of the curious historical development of the modern "Mining Camp" and its prototypes in ages past; we cannot enter upon a discussion of the various modifications of the simplest form of a Camp, ruled by the direct will of its members; with the entire criminal code, and with much of the civil code of the larger Districts, we have nothing whatever to do. It is simply the regulation of the use of land, under circumstances of peculiar difficulty, by laws mutually agreed upon, and in many cases merely verbal, that forms the subject of this investigation. Everywhere we shall find a return to primitive ideas; use is made the proof of ownership, and equality in regard to the size of the various lots is considered of prime importance.[1]

---

[1] Mr. Henry George, in his "Progress and Poverty," Book VII, Chapter V, has some interesting remarks upon this point.

We shall also discover that the legislative enactments of the Mining Camp cluster with peculiar force about this central question of land tenure. Outside of each mining district for miles extended the unsurveyed government land, like the *Folkland* of the early English. Within the District itself the territory not in actual possession of some miner was in a state analogous to the *common lands* of the village communities, and, although there was seldom any definite statement of the control of the citizens of the camp over this unused territory, yet, as a matter of fact, such control was often and successfully asserted, even to the extent of taking possession of and selling for cash all the unoccupied lots in the village. It will easily be seen that under such circumstances, extending over a number of years, a large body of laws was created, setting forth with great exactness the relations of the miner and his land claim. First, the legal size of a "claim" was determined, then the requirements of legal ownership, next, the conditions whose constant fulfilment was necessary, and the circumstances which would work the forfeiture of all right and title, and, lastly, the methods of procedure according to which disputes were to be settled.

The limit of time covered by the present investigation is the thirty-six years since 1848, for important organizations of Mining Camps differing in few particulars from those of the "flush era," have taken place within the past few months. Geographically, the field under consideration extends over the mineral belts of the Sierra Nevada and the Rocky Mountains, and includes the scattered camps that lie between. It goes as far south as the districts of Chiricahuas and Southern New Mexico; as far north as the sources of the Fraser and Saskatchewan. As regards numbers, there were in 1866 over 500 organized districts in California, 200 in Nevada, and 100 each in Arizona, Idaho, and Oregon, or a total of 1,000 small communities, each with its local laws. Since then, the number of districts has diminished in the older mining regions, and has increased in the newer ones. State and national legislation has in a great degree obviated the necessity for District law, except upon the frontier, but there the Mining Camp ideas still flourish.

The only way in which the land system of the miners can be clearly understood is by a minute analysis and close comparison of the land laws and consequent regulations of a large number of districts. We have obtained from various sources copies of the laws actually enforced for a long period of years in many of the leading camps of various States and Territories, sometimes, of necessity, abbreviating their enactments, but omitting nothing essential to a full and fair understanding of the subject. A volume of two thousand pages would hardly be sufficient to contain the complete laws of all the Mining Districts of the Far West. These laws in their complete form are usually concise, well-worded, and clear in meaning; in some cases they are evidently drawn up by lawyers, in other cases by men of good general education, but totally ignorant of law-phrases, and in a third class of cases they are the work of ignorant but practical and much-in-earnest frontiersmen. By far the greater number of districts to whose laws we shall call attention are Californian, because in that State all the essential features of the system were first developed.

The first district to which we shall ask the attention of our readers was situated five miles from Sonora, the county-seat of Tuolumne County, California,[1] and was in one of the richest gold-bearing ravines known to the pioneers of 1848 and 1849. Many nuggets of large size were taken out, and fortunes were made in a few days or hours. The homely appellation of "Jackass Gulch" was given to this district

---

[1] Our authorities for the Tuolumne Camps are in the first place, a very rare pamphlet, "Heckendorf & Wilson's Miners' Directory of Tuolumne County," Sonora, 1856; secondly, the testimony of some ten or twelve of the pioneers of that region, with whom we have corresponded; thirdly, extracts from the early Tuolumne papers, advertisements, notices of meetings, &c.

upon its first organization in the autumn of 1848. For a time it was one of the most "booming" places in the mountains, and hundreds of ardent gold seekers dwelt there. Verbal laws, agreed upon at a mass-meeting, ordained that the size of a claim should be ten feet square, as in many cases a plot of that size had yielded ten thousand dollars before being exhausted of its precious mineral, and upon that basis of allotment there was enough mining land for all the workers to obtain claims. Written laws were soon needed, and, so soon as the richest spots were exhausted, the size of the claim allowed was by common consent enlarged. In 1851, the Camp Laws, as adopted and enforced, were as follows:

First. That each person can hold "one claim by virtue of occupation," but it "must not exceed one hundred feet square."

Second. That a claim or claims if held by *purchase* "must be under a bill of sale, and certified by two disinterested persons as to the genuineness of signature and of the consideration."

Third. That "a jury of five persons shall decide any question arising under the previous article."

Fourth. That notices of claims must be posted upon the ground chosen, and must be renewed every ten days "until water to work the said claims can be had."

Fifth. That, as soon as there is a sufficiency of water for working a claim, "five days absence" from said claim, "except in case of sickness, accident or reasonable excuse," shall forfeit the property.

Sixth. "That these rules shall extend over Jackass and Soldier gulches and their tributaries."

The greatly lessened value of the mining ground in this camp is shown by the increased allotment. The requirement of claim-notice renewals during the idle season, when there was no water obtainable for washing the auriferous gravel, was common in most of the camps unless a miner lived upon his claim. In more northern camps the winter, not the summer, was the season when claims lay unworked.

In all the Mining Camps the duly accepted "claim notice" must be "good and sufficient," but if it was dated, signed, and contained the clause "in accordance with the laws of this district," it was a legal notice. Some camps required it to be "written with ink," others "painted on wood or other durable substance;" some prescribed that the "claim stakes," one at each corner of the claim, should be "four feet high and five inches square," and that at least two of them should bear "legal notices." The following examples of peculiar and amusing notices that, although legally deficient, were accepted in their time, have been furnished me by pioneers:

"*Clame Notise:* Jim Brown of Missoury takes this ground jumpers will be Shot according to the Laws of the Timbuctoo District."

Another equally sanguinary one, although not such an unconscious slander on the district laws, was couched in these terms:

"*Notis*—to all and everybody. This is my claim, fifty feet on the gulch, Cordin to Clear Creek District Law, backed up by shotgun amendments." (Signed) "THOMAS HALL."

Here are a few others:

"*Notice.*—Our claims, in this district, according to regulations."
(Signed) "PHIL. MAZEY."
"JO. HODEN."

"*Taken.*—This is my Honest Claim of ten feet each way."
(Signed) "ANDREW PESANTE."

"*To Miners.*—Look further. Respect my claim stakes driven by the rules of Douglas Bar." (Name illegible).

A claim notice of a far more definite nature than these primitive types was posted in San Andreas District, Calaveras County, in 1862, and reads as follows:

"*Notice.*—The undersigned claims this ground for mining purposes, known as the Robert McCall claim, being a deep or shaft-claim, and bounded on the northwest by the Gilchrist and Cornwell claim, and on the southeast by the Plug Ugly claim, and he intends to work it according to the laws of the San Andreas Mining-District."
(Signed) WILLIAM IRVINE."
JOHN SKOWALTER, Recorder, August 18."

A Colorado newspaper reports the following as a claim location made in 1883 in one of the Gunnison Districts:

"The undersigned claims this lede with all its driffs, spurs, angels, sinosities, etc., etc., from this staik a 100 fete in each direcshun, the same being a silver-bearing load, and warning is hereby given to awl persons to keepe away at their peril. Any person found tresspassing on this claim will be persucuted to the full extent of the law. This is no monky tale butt I will assert my rites at the pint of the sicks shuter if legally Necessary so taik head and good warnin. Accordin to law I post This Notiss.—JOHN SEARLE."

Returning to the "laws of the camps," we remark that a great number of variations upon minor points appear. The "Right to purchase other claims" allowed by Jackass Gulch was denied in many other districts. The mode of settling land disputes, also, differed much, even in contiguous districts, but arbitration was a favorite system. Springfield District, whose leaders were men from New England, trained in town-meetings and local self-government, was able to create an organic law far superior to that of Jackass Gulch, although not more than ten miles distant. We have no record nor reliable tradition of the first year of the camp except that "it was orderly." But they adopted laws in written form at "a mass-meeting of the miners April 13, 1852," and this Code was revised August 11 of that year, and again December 22, 1854.

After describing the boundaries of the district with unusual minuteness, the Preamble (of April, 1852) proceeds to declare: "That California is and shall be governed by American principles." . . . "And as Congress has made no rules and regulations for the government of the Mining Districts of the same, and as the State Legislature of California has provided by statute and accorded to the miners of the United States the right of making all laws, rules and regulations that do not conflict with the constitution and laws of California, '*in all actions respecting mining claims*,' therefore we, the miners of Springfield District, do ordain and establish the following Rules and Regulations."

Immediately following are sixteen long and precise articles referring to land rights and the settlement of disputes. The size of the claim is fixed at one hundred feet square, no person, under any circumstances, to hold more than one such claim. Upon this claim work must be performed at least one day out of three, during the entire season for mining. Each claim must have good and substantial stakes at each corner, and "must be registered and described in the book of the precinct registry," to which the owner or owners shall sign their names. Several persons, each owning one claim, may concentrate their labor upon any one claim if they deem it advantageous.

Disputes are to be referred to a Standing Committee of five miners, or to any member or members of this committee, as "arbitrators," or a "miners-jury" may be summoned. "Each member of the Standing Committee shall in each case be paid two dollars for his service." It easy to see that in many cases one arbitrator could decide a case as easily as could five, and at much less expense. The laws proceed to further define the process of arbitration. The head of a committee is to be sworn in by a justice of the peace, "provided such an officer be appointed in this mining district," and is to administer the oath to his associates, and to the witnesses. This oath was of the form: "You swear, (or affirm) to honestly and truly arbitrate, without fear or favor, between the parties in all disputes that come before you for decision." It is also declared that the verdict arrived at, whether by jury trial or by arbitration, must be received as "conclusive and binding upon the parties thereto and be deemed and considered final in all such cases." Either party may compel the other to come to trial by full board of arbitration by giving him three days' notice of time and place. Costs shall be paid in the same way as in Magistrates' Courts. Disputes over water privileges are especially named for arbitration.

The desertion of a claim for thirty days during the working season, of six or eight months, resulted in "forfeiture without remedy," and if the claim was a valuable one, some watchful person would probably take legal possession at exactly one minute past midnight on the morning of the thirty-first day. Such cases occurred very often in the mines.

Article thirteen of this Springfield District Code reads as follows:

"No person not an American citizen, or where there is any reasonable doubt of his being entitled to the privileges of an American citizen, shall be competent to act on any arbitration or trial by jury."

The next article provides that "companies which go to great expense in running tunnels" are allowed "two claims for each member of the company." The first code of "Tunnel-Claim laws" adopted in this region was several years later, January 10, 1855, and it then defined a legal tunnel claim as "one hundred feet along the base, and running from base to base through the mountain."

Article fifteen provides for the election of a district recorder "who is to have fifty cents for recording the title of each mining claim."

The last article provides that "all claims held by foreigners who have failed to secure their State license" shall be forfeited. This was a provision intended to aid in the enforcement of the State Act of April 13th, 1850, imposing a tax of twenty dollars per month upon all foreign miners.

Jamestown District, a few miles from Springfield, contained a large number of Southerners and Western men. It was an orderly and well-managed community, ruled by "Miners' Meetings" convened once in six months, with an occasional "special meeting." In 1853, "several persons having attempted to pass unpopular laws," the miners held a rousing assembly, repealed "all previous laws of every sort whatever," enlarged the bounds of their district so as to include a number of outlying claims, adopted the usual standard size of one hundred feet square "in place of the previous and varying regulations," and declared that "all claims secured under former laws" were publicly acknowledged as legal.

Within three days after the time of location a claim must have a ditch one foot wide and one foot deep cut about it; notices must also be posted, and stakes driven at the corners. Failure to work a claim within six days after the mining season begins causes forfeiture. A miner can hold other claims only upon proof of purchase. Miners shall have the use of water from the ditches "according to the date and situation of the location of their claims."

An important and very significant clause is to the effect that "miners may dig up any farm, or enter within any enclosure," by giving the owner security "that they will pay all damages inflicted." In no case, however, shall they dig "within twelve feet of a building, or obstruct the entrance." Mining was held to be the paramount industry, and the miner's possessory claim outranked that of the agriculturist. Payment of damages meant only that growing crops and improvements should be compensated for if destroyed.

When this camp was first established, in August, 1848, its wealth was so abundant that a trader sold a handful of glass beads to an Indian for about $6,oco in gold-dust; and for some time in this and adjacent camps men wandered over the hill-slopes plucking up tufts of the coarse "bunch-grass" and shaking off the soil in buckets, often thus uncovering rich "pockets." The same thing was done in Australia and Brazil in the early days of the "placers" and the "dry diggings." The irregularity with which gold is distributed, even in rich districts, makes the dignity and minute enforcement of the camp land laws all the more remarkable as a triumph of the Anglo-Saxon capacity for self-rule. Hundreds and thousands of times, in camp after camp, "ill-luck" seemed to follow one miner, while "good-luck" was another's constant attendant. One old pioneer writes me from Idaho, after a quarter of a century spent in placer camps, that he has "tossed up" with a miner for choice of ground, the two having arrived on the spot at the same time, and has chosen the upper claim, and "failed to make wages," while the other man "took out five thousand dollars from a space twenty feet square and four or five feet deep." Of course there was claim jumping at times. All laws are occasionally broken by unruly members of the community, but an impartial historian of the Mining Camps is forced to declare that these infringements of individual rights were in most cases punished with terrible severity by the citizens. The famous "Holden Garden Claim" fight, near Sonora, Tuolumne County, about 1852, is a case in point. A very rich ledge of gold-bearing quartz having been found by some gardeners in their "cabbage patch," was "staked out" by them, and work begun. A party of gamblers in the town started for the place, and attempted to take possession—did in fact hold the ground for a few hours, but the citizens took a hand, drove them off, and restored the property to its rightful owners, after a skirmish in which one or two men were killed and several wounded.

To return to our synopsis of the special laws of various camps concerning land and land matters. We next take up a small but very active California district, "Shaw's Flat." In this camp there had been much trouble from the attempt of several miners to run the lines of their claims so irregularly as to include more of the good mining ground than they were entitled to. This was an early difficulty in hundreds of camps, human greed and selfishness being distributed with considerable evenness, in camps, as in cities. Men tried to establish the legal idea of a right to forbid others from trespassing on their "claims." This failed utterly; in every camp, without exception, the fact that a prospector had occupied the mouth of a gulch or cañon gave no right whatever to forbid others from "highway privileges" across his claim. The miners of the Sierra would have laughed to scorn any such right. Pastoral land claimants in the West have often monopolized and controlled thousands of acres of land by the mere location of a "warrant" on a section that contained the springs, or was the gateway of a precipice-guarded valley. But never, in all the mining history of the Pacific Coast and

Rocky Mountain region, did any prospectors succeed in this fraudulent extension of the "possessory right" to use land. "Shaw's Flat," and other districts settled the difficulty by ordaining "right of way" everywhere; and by insisting upon the "square location," except in a few well-defined situations.

We are speaking, it must be remembered, of placer mines, not quartz-leads, to whose holding a somewhat different law, evolved partly from the placer usages, and partly adopted from the experience of other races, was applied. Quartz-ledges in the United States follow this oblique location theory; the ledge itself is the property and may be followed to any depth. In Mexico, the square location system is the one adopted for quartz-mines as well as for placers. The true placer claim is so much surface, and to bed-rock, or beyond as far as crevices that may contain gold can be found to extend. We find, therefore the usual clause "in one lot, and square in form," which prevents the prospector from conforming his claim to the outline of the gulch. Sometimes, when the width of the stream is uniform and the sides of the ravine nearly parallel, the clause reads "shall be uniform in width, and extend from bluff to bluff," or, "shall run across the width of the gulch." So narrow and deep are most of the wild mountain gorges in which the miners toiled, that it appears likely that the typical camp was of this latter form—a long line of men extended for several miles up and down the ravine, and returning at night across each other's claims, to the little collection of tents and cabins, saloons and hotels, and motley adventurers, that was the "Camp," the district town, the temporary metropolis, perhaps, of half a dozen surrounding camps, but doomed to sudden downfall the moment the gulch was "mined out," or even long before, if a richer place was discovered elsewhere.

At "Shaw's Flat," a "legal claim notice" was sufficient to hold a placer 40 feet square for ten days after work was begun in the district, counting from midnight. "Part of a company" could not presume to "hold the claim of a whole company during the absence of any of its members." There were to be no non-resident stockholders, no taking up of claims in the name of distant friends and relatives. If a man left a district for the working season he must sell his claim to some one who would utilize it. The "Camp law" held that a "company" was but an association of men whose capital was their own work, and who found that their claims could be handled better thus. The development of this idea in a few years, in the mines, to a point when associated capital constructed some of the most remarkable of engineering works, and brought water for hydraulic purposes ten, twenty and forty miles over the most broken and difficult of regions, would of itself form a fine theme for an extended essay, but the subject is foreign to our present purpose; we have only to deal with the primitive forms of the "company" idea in the mines, and in most of these no truly associative and representative element was recognized as inherent in a "company." Some interesting exceptions, however, have been noted, and we shall call attention to others in the course of this investigation.

The laws of "Shaw's Flat" proceed, further, to provide that all "deep diggings," where "pay-dirt" is 25 feet from the surface can be "laid over without work from December 1st to May 1st, if they are only well defined by marks and stakes, so that no difficulty need arise," and also "are recorded in the district register, which shall always be open to inspection." The point involved here is the desirability of giving constant employment to every citizen of the district; "surface," or "flat," or "river," or "bar" claims were worked when water was abundant; deep claims were "drifted," and the "pay-dirt" carried down to the streams or springs, and then washed out, for which there was sufficient water even in summer. Sometimes they could be worked in winter too, but no one wished to do that; for he wanted some occupation when the water ran low in the channels, and the rockers could no longer be swung. In this district the annual meeting was attended "by all claim owners" unless they made "a reasonable excuse." This reminds one of the seventeenth century local laws of New England towns. In Farmington, Connecticut, for instance, whose old town records the writer had occasion to study the last summer, it was ordained, two centuries ago and more, that whosoever failed to attend the annual town-meeting, on him a "fine of twelve pence" should be laid, no light matter in those days. And in the ten or twelve California districts of 1849--54, wherein we discover special clauses intended to strongly enforce the duty of each and every miner to attend the annual and semi-annual meeting, the New England influence is apparent, the names of prominent men in each camp are New England names, or in some phrase of the enactment itself the link is clearly shown. For instance, there is a forgotten camp near Piety Hill, Shasta County, long ago swept from existence, not a house, or cabin, or tent-pole, or wheelbarrow left to mark the spot, only vast gravel mounds, vast heaps of hand-piled bowlders, vast cuts in red clay, now overgrown with pines. Here the "Camp" had a "committee man," the head of a committee of five, to settle disputes, record claims, and preside over special meetings.

Four small districts of 1850, "Saw Mill Flat," "Brown's Flat," "Mormon Gulch" and "Tuttletown," had laws that were much alike, and at one time there was talk of uniting them into a "confederacy," together with several adjacent camps, all in Tuolumne County. That is to say, tradition and the memory of old settlers report that the scheme went so far that delegates met and talked the subject over. They reported, "There's no money in the plan; if we wanted to dig a ditch in common we could easily unite for that, but our laws are sufficiently uniform to prevent annoyance." Two of these districts begin their laws with "Whereas this district is deficient in mining laws and regulations, and disputes have arisen, therefore we, the miners of ——— District, in convention assembled, do pledge ourselves to abide by the following laws." "New diggings" in the vicinity of an established camp are always much desired, so we find here, as in some instances previously noted, that the discoverer of such is rewarded with a double claim. The Carthaginians used to build temples to those who found new mines, and paid them honors after death, but the rightfulness of giving to such a double share in the mining ground was not recognized, so far as we can discover, until men of our Germanic race began to organize into mining communities. Some few of the California Camps felt that a treble share of the gold-streak was none too much for the discoverer, and one instance of a larger allotment has fallen under my observation, that of "Poverty Hill," where four times the usual claim was allowed.

"Saw Mill Flat," when first organized, about 1850, provided for a committee of three persons, chosen by the miners in general assembly, to "see that the laws are obeyed," and to "call meetings of the miners of this precinct to enforce the laws, or whenever for any reason they deem such a meeting necessary." The arrangements made for "arbitration" in this district were minute and definite. "Whenever any dispute shall arise respecting claims or water privileges each party shall choose two disinterested persons, the four thus selected shall choose a fifth, and the five thus selected shall then hear evidence, according to the laws of the precinct." The law of "Brown's Flat" ordains that these five arbitrators, instead of being chosen by the disputants in the above manner, shall simply be "appointed, whenever needed, by the committee" of three duly elected persons who ruled the Camp. And they were especially enjoined to "view all disputed territory," and to "summon and examine witnesses" when necessary. This "Brown's Flat" committee was chosen in public meetings "to hold office until superseded." Its powers and privileges were about the same as at Saw Mill Flat. Sometimes the arbitrators failed to give satisfaction in these camps, or found the case on which they were called to act so difficult and obscure that they retired promptly. What then? In less highly organized camps, still governed by successive impulses of popular will, a meeting would be called, and the case stated, *pro* and *con*, to the assembled claim owners; but camps where "committees" ruled made these committees the "Courts of Appeal," and if arbitrators retired, or if their decision was disputed, the "Committee of the Camp" was called upon. No case of a still further appeal to the State or County Courts, then in existence, can be obtained. The final decision came from the camp authorities. Nor were these decisions matters of small account, for property worth several thousand dollars was often involved, and its ownership was settled in one afternoon's visit from the arbitrators; to such an informal affair had the administration of justice been reduced.

The Tuttletown laws say that "no person shall ever hold more than two claims in this district, either by purchase or otherwise;" again, an admission of the plan of having separate claims for summer and winter. They also provide, with much circumlocution and elaborateness of detail, for the marking of each claim, and ordain that "any one who destroys a notice or claim stake shall be fined not less than five dollars nor more than fifty dollars." To this some of the northern camps added, "and upon a repetition of the offence, shall be requested to leave the district;" of course, in that case, forfeiting his claim. Tuttletown required that written notices of claim locations should not only be placed on the ground itself, but must also be posted "in some convenient and public place in the district." This place in nearly every case was the door of the "leading saloon" of the town, or the trunk of some large tree in front of that building. In "Hay Fork," Trinity County, in 1879, also in "Middletown," in the same county, and in "Ophir," Placer County, in 1881, the writer saw "notices" of placer and of quartz locations posted in this manner, signed by the deputy of the County Recorder, and "in accordance with the customs of this district." A common way of putting up a claim notice is to burn it with a red-hot iron upon a slab of pine or cedar, preferably the latter, and if this be then nailed with blacksmith-wrought nails to a tree, it withstands many seasons of sun and storm. Such "claim writs" yet remain in some of the California cañons; saloons where hundreds congregated have mouldered into dust, and over the rude chimneys of miners' cabins, blackberry and clematis vines tangle, pink wild roses and scarlet mimulus blossoms glow; while the scorched bit of cedar shingle is the most conspicuous evidence of the stormy past, when this wild ravine, now a solitude, held a frontier settlement of perhaps a thousand able-bodied miners, who partitioned out every foot of its soil.

Three other districts of which we have obtained copies of the early laws made by pioneers were "Poverty Hill," "Chili Gulch" and "Yorktown," all of them organized early in that "golden prime of Forty-Nine," of which the mining ballads of the Far West speak with such rapture. If it were part of our present task to deal with the general institutional relations, history and literature of mining camps, the little group of a dozen such settlements in Central Tuolumne would afford materials for many chapters. Here, within a radius of ten miles, were camps differing from each other in many important respects, as regarded civil code, criminal code, character of population and form of government. It is easy to believe that if State life had been delayed but a few more years, if territorial forms had prevailed for a time over the broad California gold fields, these highly vitalized communities would have crystallized into much more permanent units. Even as it was, the uncertainty of mining life, the sudden downfall of flourishing camps, was the cause, far more than State organization, which prevented the growth of permanent geographical and political divisions other than townships, within the county. "Tuttletown," and "Brown's Flat," as we have seen, had their "committee rule;" but "Chili Gulch,"

"Yorktown" and "Poverty Hill" went to old Spanish and Mexican law, and gave obedience to elected Alcaldes, whose rule was almost absolute, and from whose decisions there was no appeal except by revolution. Here were camps with but a ridge of quartz-ribbed, pine-crested hills between—and while some looked to New England and the depths of the Odenwald and the Black Forest for the source of their scheme for settling land disputes, others, equally well governed, looked to sources Castilian, Moorish and Arabian. Yet it is a matter of history that Americans ruled in all alike, though adopting different methods, governing "Chili Gulch" as fully as "Tuttletown."

The laws of "Poverty Hill" limited deep diggings to claims of thirty feet square on new ground, and of fifty feet on ground which had been once worked. A claim must be occupied and fairly opened for mining operations within three days after posting the notice, and absence from the claim for a period of ten days during the working season "throws it open to re-location as an abandoned claim." This camp was "named by contrary," for it was a remarkably rich gulch, and its "claims" were seldom or never forfeited by the fortunate possessors. The practical working of Mining Camp land laws in reference to the forfeiture of "claims" can easily be understood. They were rude tests of judgment and perseverance. Claims that afterwards proved the most valuable on a stream were often located, abandoned, re-located and again abandoned, before any one came along who had the "sand," to use a peculiar Westernism implying "grit," and who stayed by his work until "bed rock" and a rich reward were reached. Good and successful miners have told me that their rule was to try every abandoned claim and follow their own individual opinion about the "prospect," regardless of the fact that the expired "notices" of a dozen former owners might be scattered over it.

Two other districts that presented interesting features were "Gold Springs" and "Chinese Camp." The former "earnestly recommended arbitrators," but did not consider them "essential," as a camp jury "might be called." It required all miners to pay strict regard to the condition of roads in the district, also to horse-back trails, accustomed foot-paths, and "not to destroy them in mining operations" without replacing the same by new ones. In other words, if the people of the district were using a path, or street, or roadway across a miner's claim, he must not undermine or destroy it without preparing another on the hillside, or in some convenient way around the place of operations. This principle, also, forms an extensive subject for legislation in the early camps, and takes varied forms. The difficulty was to enforce it; men would wash a road-bed into the creek, and then protest that they had not made a cent, and perhaps move out of the district before the Committee or Alcalde could serve a legal notice upon them. Some later camps obviated part of this difficulty by enacting that the new road "should be made ready for use according to public satisfaction, before work was commenced on the claim crossed by the old road." The sites of many of the towns in the mines have been "drifted" from end to end; towns have been moved, and every foot of the soil on which they stood "staked off" and mined; streets have been pre-empted, and highways used for years have been made "mining property" by the accident of a sudden pick blow or land slip.

"Chinese Camp" allowed its chief officer the munificent reward of three dollars for each decision in a land dispute, and also his mileage of one dollar per mile for the distance from the central point of the camp to the disputed claims and return. There was a time in some of the earliest camps when fees were much higher, amounting in some cases to one ounce of gold dust for each case tried, and even to two ounces (thirty-two dollars), but mileage was nowhere adopted until after 1858, and by that time one or two dollars was considered an ample fee in most cases.

There was no more characteristic district in California than that of Columbia, in Tuolumne, and its complete laws will be found in the Appendix. This district included "Yankee Hill" and several other lesser camps, at first separate, so that it presents an example of union. March 27th, 1850, five New England prospectors, three of them from the woods of Maine, discovered the famous "Kennebec Hill" placer deposits, and within thirteen days there were eight thousand miners in the new camp, saloons, hotels, stores, and all sorts of human parasites of the actual workers of this mushroom city. Fourteen days from the day of discovery a meeting was called "to organize and govern this camp." "Alcalde," "sheriff," "register of claims," were the officers, and the fees collected by the latter paid all expenses of the organization. In 1852, the united district polled more than twelve hundred votes. The laws adopted were enforced and the place was noted for its orderly appearance.

A very interesting feature of district organization is shown in the course sometimes taken when the surface mines began to fail and more water was needed, through "mining ditches." Associated capital constructed these at a later period, but several districts in the southern mines prolonged their existence by the simple process of forming a "work-stock" company in which there were as many shares as the estimated number of days' labor required to construct the needed aqueduct. Then it became a matter of local pride to do one's share, and the several lesser camps along the line of the proposed ditch sent out all the men that could be spared to take part in the work. Details are unfortunately lacking, but the general system was as outlined. In 1855, several hundred miners from Columbia aided in the construction of such a ditch.

Little Montezuma camp kept apart from its neighbors, and allowed each miner to locate "three squares of one hundred feet each," together or separate, as a "surface claim." Its regulation Tunnel Claim was one hundred and fifty feet in width, and of any depth desired. Its "deep sinking" or upland claim was one hundred by three hundred feet; in other words, it differed from the surface claim in only one particular—that the "three squares" must be contiguous. One week was allowed for recording, and three days' work each week was essential to the end of the chapter. The "Recorder" may be at any time deposed "by a two-thirds vote of the resident miners of the district." Curiously enough, he is also "Arbitrator-in-Chief" presiding over the "Arbitration Court" of four members, two chosen by each disputant, and he casts the deciding vote in case of their failure to agree. Fees as recorder, one dollar per entry; as arbitrator, one dollar per case; cost of a decision to two disputants, two dollars and a half apiece, and no more—this in gold-dust, weighed out at the valuation of $16 per ounce. If one of the disputants "backs out," refusing to select his men, or refuses to abide by the decision, the Recorder is empowered by the district to declare and enforce his judgment as final. Of course he can only appeal to public opinion, call a meeting, and ask for endorsement and deputies. But to this complexion matters never came. The Recorder ruled in peace so long as the camp had any existence.

As late as 1856, the thriving "French Camp" district in Stanislaus County provided for three, or five, or seven arbitrators, as the disputants chose, and added this Spartan mandate: "In the event of any parties not acknowledging the decision, then the miners of this district shall meet and compel said party to recognize the umpire's decision."

Sweetland Mining District, Nevada County, was organized in 1850, claims then being thirty feet square; two years later these proved too small, most of the valuable ground being exhausted, and, after many debates, the size of claims was fixed at 80 by 180 feet. It is impossible to ascertain why this particular size was chosen, unless the explanation that one pioneer gives, that this was about the equivalent in value of the earlier allotment, be the correct view. This was in 1852, and the next year the miners met and decided, because of local differences and disagreements, to divide the district into three, each of which thereafter made its own laws. North San Juan was one of these districts, and has been famous ever since for its enormous hydraulic operations. In 1880, the writer spent several weeks along the "San Juan Ridge," visiting the dozen or more small towns where mining was being carried on, and where, although great corporations held most of the valuable ground, much of the old "district law" still survived, and picturesque tales of the past were abundant and realistic in the extreme. The early codes of this region, now only existing in tradition, allowed but one claim to each miner; at a little later period "one claim by right of location," and an unlimited number by purchase. The "notice" must be renewed "every thirty days," unless this necessity is obviated "by the daily presence of the owners or their proper representatives." An expenditure of $500 in prospecting into, or opening up a claim, is counted as possessory work for two years. The quartz laws of this District, and others in Nevada, merged into the county "quartz code" adopted at a general convention of the miners.

The earliest placer claims in Nevada County were ten feet square, a few weeks later they were increased in size to "fifteen feet on the river," although the size soon varied even in adjoining camps. At "Brush Creek," whose placers yielded over three million dollars, the size allowed was "a square of sixty feet each way." A chapter could easily be devoted to the early history of the famous camps clustered about Grass Valley and Nevada City. The late Benjamin P. Avery, a brilliant, earnest and much-loved man, whose literary powers were late in ripening, and whose life, by reason of its large plans, was pathetically unfinished and uncrowned, wrote a letter, years ago, to the editor of a little Nevada County Directory, giving a graphic account of the early history of that famous network of camps. It was in October, 1850, that he started from Mormon Island, now Sacramento County, on a prospecting tour to Redding Springs, now Shasta City, and several hundred miles distant. He "rode a little white mule; pork, beans, hard bread, and blankets packed behind." Hearing of "pound diggings," or those yielding some $200 per day, he changed his course to Gold Run. At that time Caldwell's Upper Store, now Nevada City, flapped its canvas sides, and protected, to the best of its ability, a slender stock of dollar-a-drink whiskey and dollar-a-pound flour and biscuits. Down on the flat were a few tents, and the bars were being worked with dug-out cradles and wire and raw-hide hoppers. Pork was $2 a pound and boots cost $80 or $90 a pair. Mr. Avery found good diggings, and returned for his companions, but when they arrived the entire gulch was "occupied by long-haired Missourians, who had staked out their thirty-foot claims, and were taking out their piles. At night many a long-tom party took a quart tin pail full of gold to their cabins."

Those were the times when it cost two dollars and a half to have a letter carried from Sacramento to the camps, when Mrs. Stamps, the wife of the first elected Alcalde of the region, and her sister were the only ladies in the county, and when lawlessness was quickly suppressed and the steady increase of social and protective organization was everywhere manifest. The first Alcalde of Nevada City was elected by a voting population of 250, but in many camps ten or a dozen men chose this peculiar and all-powerful officer, giving him all the powers granted under the Mexican and early Spanish system. He became the judge of the village, the petty lord of the tented town, and only the voice of the people could bring his powers to an end.

The flush mining camps have often been described, with their curious Sabbath-day mingling of ministers, gamblers, auctioneers, dog fights, and street sales, all concentrating about the gorgeous saloons, where monte, faro, roulette, poker, vingt-et-un, and other games of chance were in full blast. There were Indians, Mexicans, Chilians, Hawaiians, Asiatics, Europeans, Yankees, Westerners, Southerners, men fresh from their claims, still begrimed with auriferous mud;

men dressed in the latest fashions of Paris, each one of them all measured in that virile, sinewy community for exactly his worth of manhood. But tales of daily life in the camps are not so frequent.

There is a story told about Main street, in Nevada City, California. In 1851 some miners began to sink a shaft in the middle of the street, and in the most important business centre of the town. Expostulations, long continued, were of no avail. "Miners' rights come first," the intruders said, "and there is no law 'gainst diggin' in the street, an' we mean to dig." The storekeeper whose property was most in peril went into his store and returned with a loaded and cocked revolver, which he pointed grimly at the miners, already neck-deep in their shaft.

"Then I'll make a law," he cried. "Just you boys go back and hunt up a rich gulch. No gold here; plenty of lead."

"That's good law, Judge," responded the leader. "Better'n the average Supreme Court decision. Boys, let's fill up the hole and start for Last Chance or Timbuctoo." A later tradition reports that the party "struck it rich" a month later, and sent a nugget in remembrance to the irate citizen who drove them from Main street.

Gold Flat, also in Nevada County, was rather a typical camp. In August, 1850, only two cabins were on the ravine; in July, 1851, over 300 miners were living there; in 1852, it was "played out" and abandoned. Moore's Flat, discovered in 1851, held 500 miners before the close of the year. Orleans Flat within a year after its settlement boasted of several hotels and a voting population of 600. Grass Valley itself, which in October, 1850, had but fifteen cabins in sight, was multiplied more than tenfold within five months, and in five years had a population of over 3,500. Little York, prospected in 1849, had its "flush times" in 1852, when "pay dirt" that "ran a dollar to the pan" was discovered. The miners met, laid out a street, declared it sacred to town purposes, and ordained that each claim owner should be entitled to a building lot in the town. Washington Camp, settled late in 1849, contained 1,000 miners by August, 1850, and before January, 1851, there were more than 3,000 miners in that region, governed by laws of their own devising, and divided into several districts.

Rough-and-Ready, another celebrated mining camp, had most of its auriferous gravel claimed by two "companies" who discovered the placer, organized, and tried to exercise rights over a large tract. They sent to the Atlantic States to hire workmen under contract for a year. Long before these men arrived, hundreds of "prospectors" had calmly and peaceably taken possession, and divided up the gravel beds according to its usual camp regulations. Without a shot being fired, the entire monopoly scheme vanished into thin air, and the twenty original discoverers considered themselves lucky to be allowed to hold claims of the same size as those of later arrivals. At a little later period in this camp the process of consolidation of claims began with great energy, and about thirty "companies" were thus formed to work the deeper deposits. One of these "associations" of ten persons took out $400,000 in two seasons, from their ten claims, worked as one.

Among other early camps of Nevada County whose land laws were strict, and well enforced, were Keno, Boulder Bar, Brass Wire Bar, Lizard Flat and Poorman's Creek. The entire gold product of the Nevada Basin, which surrounds Grass Valley, during this era was not less than $3c,000,000, and every dollar of it was taken from claims whose possession was regulated by camp law and public opinion, and by no outside force whatever.

Coke's "Ride over the Rocky Mountains" (London, 1852), contains some good illustrations of early district law. In the Mariposa Mines, which the author visited, the local enactments made the size of the claim thirty by one hundred feet, and claims of this size changed hands frequently, as high as $2,000 being sometimes paid for one such "possessory right." "Societies" had been organized in several cases to work mineral ground in common, and a number of quartz mines were in operation. The camps were orderly and well-governed.

Another English observer[1] gives a vivid description of life in the mining regions. He speaks of "Murderer's Bar," the swift river, the black, obstructing rocks, the village-tents of sun-bleached drilling, the miners waist-deep toiling to turn the course of the river, others in pits, digging to bed rock, some alone, some in companies; all life, vigor and determination. He says: "Every digging has its fixed rules and by-laws. All disputes are submitted to a jury of resident miners. In certain cases twenty men or so from one camp are met by a number from another camp." He goes on to say that disputes sometimes arose, and even demonstrations with fire-arms, but that good sense always prevailed. To this endorsement, he adds: "I have had my placer claim of ten feet square encroached on, I appealed to the crowd, and a committee of three being at once chosen, measured it from my stake, and being found correct, the jumper was ordered to confine himself to his own territory, which he always did."

The number of mining districts which organized at an early day, under definite regulations and arbitrators or committees for settling disputes, was very great. There were Jackson, Sutter Creek, and Volcano, in Amador; Placerville, Mud Springs (which as late as 1863 had provisions for arbitrators), Georgetown and others in El Dorado, Cherokee, Nimshew, Hungarian Hill and others in the Northern Mines. Bangor, in Butte County, allowed one hundred feet in width, "from rim to rim" of the deposit (1862). Forbestown, in the same county, allowed ravine claims one hundred feet wide, extending across the gulch, and required two days

[1] Mr. Frank Marryat: "Mountains and Molehills." London, 1855. (Illustrated with great vivacity by most amusing sketches of Mr. Marryat.)

of work per month (1863). An adjoining District only allowed claims of half that size.

Copper Cañon District in Calaveras County, organized in 1860, required claim stakes at each end and claim notices renewed yearly; one day's work each month; visit from the recorder and registry essential to perfect title; arbitration a law of the camp. Another Calaveras camp, "Pilot Hill," after passing through the earlier stages of small plots of land, settled upon fifty feet by one hundred and fifty feet as the proper size for "gulch claims"; two hundred by one hundred feet for "surface claims"; and for "each tunnel or shaft claim," one hundred feet in frontage, extending through the hill. Upon the last class of claims work to the value of $25 per week, as decided by the personal inspection of the Recorder, must be accomplished by each company. "Occupation and use" is sufficient to give possessory rights to the other classes of claims.

New Kanaka Camp, Tuolumne County, affords another instance of a personal inspection by the recorder of the work done, to see if it were sufficient in quantity and quality. There can be no doubt of the weakness of many camp regulations in this respect. The practical safety in thriving camps was that non-compliance with the laws might end in the entry of a claim jumper with the plea: "Forfeited and open to preemption."

In some small camps near Pilot Hill the practice was to mark claims not merely by "notices," but also by stamped tin tags, put on by the recorder "in the presence of witnesses," and bearing the owner's name and the date of his preemption.

The Brown's Valley Camp, a busy settlement in Yuba County, now chiefly an agricultural and fruit-growing district, presents some unique features in its system. The size of claims, the amount of work to be done, and the general form of government do not differ materially from those of adjacent camps, but on several occasions the miners appear to have had "political struggles," and laws that one party passed when in power were repealed by the other at a later date. In 1853, therefore, the camp resolved: "That each claim shall be entitled to one vote in the miners' meetings of this district, by the proper owner, or may be represented by a power of attorney from the proper owner." The nature of this power is closely defined, and limited to specific directions on definite subjects of dispute. It is further declared that claims must be forfeited if they fail of representation at the semi-annual meetings. For more than ten years this law was enforced, the only case, so far as I am aware, of such a regulation, and in this case due to causes entirely local.

In addition to the numerous camps of importance and some degree of permanence, there were a countless number of evanescent camps, too temporary for any particular organization, except the occasional camp convention, or perhaps not even that, if all went in an orderly manner. If a few prospectors found a small deposit of mineral they began to work it with but one idea in mind, that of obtaining as much as possible before any one else discovered their new camp. When this happened, the new comers suggested some definite size for claims, and by "mutual consent" and desire to compromise they soon arrived at some understanding. In this primitive, informal way many a group of men worked together for several weeks or months, in a mountain gorge, and practically made an even division of the gold there. In several cases on record the first workers laid out unusually large claims, to which later groups of miners demurred, and the question was put to vote; thus, in one instance, reducing each of the eight original claims to one-fifth of its former size, but without any objection from the owners, who, even then, retained larger claims than later arrivals were able to secure.

Hundreds of large and small camps, such as these which have been described, were dotted over the mining counties of California, and the local land laws of hardly two of them agreed in every particular. The title they gave was, however, a complete one for all practical purposes, and many thousands of dollars were paid for "possessory rights," and the value of such "claims" was adjudicated by State Courts in a multitude of decisions of the highest importance in the complex history of mining jurisprudence. The "local law" became an accepted legal fact, upon which the ownership of placer claims, surface or deep, tunnel rights, water privileges and quartz ledges was absolutely dependent. The local land laws of the camps may therefore be traced through the long contests in the higher courts when witnesses were summoned to give the "common custom and usage" of the district; but no ordinary law report gives these points, and we must depend upon the meagre details in the newspapers of the period. Even from such sources the total mass of material relating to camp transactions becomes large enough for a volume.

Almost every one of the early mining camps was swept by flames at various times during its existence, and to this fact the paucity of records concerning the exact wording of the first regulations of camps that have since become large and important towns is justly attributed. The wonder is that so many have been preserved in all their details, so that complete examples of all the leading forms of camp government can be found.

The laws which provided for the acquisition and continued possession of quartz lodes, leads, or ledges, were developed from the placer-miner regulations, and varied quite as much. Under their protection, however, extensive companies were formed, and costly improvements undertaken, tunnels driven many feet through solid rock, shafts sunk on "quartz-claims," expensive roads constructed, and mills of large size, fitted up with the latest machinery, were built, the value of the entire "plant" being dependent upon the local law of the camp or district. In many parts of California the quartz laws remained at this district stage of development until the habit of entering all valuable mines as United States Mineral Land, subject to purchase, and registered in the land-office, gave the local recorders little to do. But in the four important counties of Nevada, Sierra, Tuolumne, and Sacramento the need of more complete organization and greater uniformity than the district plan afforded was felt at an early period.

The miners of Nevada County assembled in November, 1852, at which time there was a great excitement throughout that region on the subject of quartz mines. They discussed the subject in all its bearings, and decided to unify and simplify the codes of the separate districts bearing upon the holding of quartz claims; the desire to promote convenience and to encourage the advent of capital to further develop the mineral resources of the country were given as the urgent reasons for this step. They appointed a committee of practical miners to examine into the various rules and regulations in use elsewhere and report at another meeting. The convention then adjourned until December 20th, when it reassembled, adopted, after full discussion, with some changes of minor importance, the code of quartz laws presented for their approval, giving it jurisdiction over all districts within the county. This code remained in full operation in 1880, and at that time "had never been changed or abrogated," and it still remains as a basis for ownership of all unpatented quartz claims.

Sierra and Sacramento counties were later in their adoption of a general system, the former in 1855, the latter in 1857. The miners of Tuolumne County assembled in 1858, and organized in the same manner. Similar meetings were held in several other counties. The size of claim allowed under these county codes varied greatly; in Nevada it was "one hundred feet upon the ledge with all dips, spurs, and angles;" in Sierra it was "two hundred feet on the lode with a width of two hundred and fifty feet on each side thereof," or a total width of five hundred feet; in other counties three hundred feet in length was not an uncommon regulation. The requirements in regard to work done, and manner of holding and registering a quartz claim, also varied greatly in different counties, but the codes in this respect were based upon a greater regard for permanence than were the placer-camp local enactments, and they always provided for a full and complete title in the end, subject to no further work tax, and possessing a recognized commercial value everywhere. Advertisements of the time often read: "Requirement work completed; claim perfected according to District Law."

Innumerable instances of separate District quartz laws might be given; the county organization was by no means the rule. Placerville District, Eldorado County, adopted quartz regulations March 21, 1863, providing that each claimant "may hold two hundred feet on the ledge or lode, with all its dips, spurs and angles, and two hundred and fifty feet upon each side thereof." Seven days' work each month is required. In 1866, the number of recorded claims was 186. Mud Springs District in the same county adopted regulations April 7th, 1863, and allowed prospectors to hold 300 feet on the ledge and 300 feet on each side. The claim notice holds it for 20 days, and the recording of the same holds it for 90 days, before the expiration of which time labor to the amount of $250 must be spent upon actual development of the mine. Provisions are made for a board of three arbitrators. The regulations of Georgetown District, same county, adopted December 10th, 1866, allow a claim of 200 feet on the lode, and require the recorder to visit, measure and stake each claim, receiving fifty cents for the service, from each locator thereof. The expenditure of $500 in labor upon a claim shall hold it for two years, without further work.

Meanwhile a large number of important decisions of State Courts between 1851--1865 had shaped the varying local law relative to "claims," "forfeiture," "labor requirements," "ditches," "flumes," "water rights," and similar subjects, into a system of more universal application, and this process of judicial interference was destined to continue. Yet the very existence of the free miners' code had been in danger during the early days of California. When John C. Fremont was State Senator he introduced a bill to establish police regulations throughout the entire mining region, and levy a small ground tax upon all miners. At that time—the winter of 1850--1851—such a measure would have destroyed the individuality of the local institutions just then rapidly developing in the energetic camps of the gold-seekers. A violent opposition to Fremont's plan was at once manifested; the miners never forgot his ill-timed suggestion. The discussion which followed in the Senate was long and excited. Many of the Senators favored a sale of western mineral lands in large tracts to the highest bidders. But it is chiefly to Senators Seward and Benton that the tacit acceptance by the nation of the policy of free mining is clearly due. They insisted on delay, urging that haste was unnecessary, and ill-suited to the preponderant importance of the subject. A year later the local regulations of the Mining Camps had assumed such satisfactory control that further and further congressional delay appeared amply justified. Meanwhile, in 1851, the Legislature of California made its declaration after investigating the entire subject, accepted the rules and usages of the miners as evidence in all controversies, and yielded up, for a time, a part of its jurisdiction. The refusal of Congress to pass mining laws being accepted as a practical recognition of the principles for which the miners contended, they pushed forward with new zeal their daring explorations, their costly investments in machinery, in water ditches and tunnels, and other needs of their gigantic undertakings.

Placer mining was in its decadence in most of the Sierra counties of California by 1856, but the chief features of district law continued much longer. Claims were so held until 1868 in parts of Lassen, Sierra, Nevada, Placer, Yuba, Butte, Shasta, Trinity, Calaveras, Mariposa, Sacramento, El Dorado, and several other counties, and over a territory several hundred miles in length and thirty or forty miles in width, the true auriferous belt of California. In many, perhaps in all, of the then existing placer camps, these laws were in force so long as "surface diggings" lasted. When the placers were exhausted and deep mining took its place, the nature of the laws was essentially modified, and the distinctive features of the district organization were necessarily modified, in some cases entirely destroyed, long before the general Government passed its first Act relating to these mineral lands (1866). But, throughout the old mining region, whereever "surface claims" still remain, they are worked, or abandoned claims are "taken up" according to old District laws of the region. Quartz claims are still worked and held as simple possessory rights, until it seems desirable to claim and "enter" them under United States Law.

But, as the opportunity for the full enforcement of such laws passed away in California, those laws conquered new territory, and spread over more extensive areas than even that great Sierra Nevada land, furrowed by the channels of ancient pliocene rivers, and sown with flakes and nuggets of the precious metal. Old California miners introduced these "laws of the camp" into Oregon, Nevada, Arizona, Idaho and British Columbia. They sailed up the broad and mysterious Yukon, and, at the present time, are ruling, with firm and equal law, the camps of Alaska. They re-organized the mining districts of New Mexico, where, as long ago as 1830, placers had been mined by the Spaniards and Mexicans, each worker being allowed to claim a circle of ten paces in all directions from his pit, and only a specified amount of work serving to hold the "*tabor*," or claim, against rivals.[1] They even organized, under their free mining system, in South America, and the Central American States. Thus the most valuable features of this local land law were widely adopted in frontier society, and still exist in camps of the Rocky Mountains.

This gradual extension over so great a territory was the work of a number of years, greatly accelerated by occasional

[1] Prince: "History of New Mexico," p. 242.

"stampedes" or "mining excitements," sometimes quartz, sometimes "placer diggings," but always resulting in exploration, and social organization of pioneer communities. It was natural that these prospectors from California should adopt to a great extent the regulations that California experience had shown desirable, with such local modifications as the nature of the locality, the water supply, the richness of the placers, or the character of the ledges seemed to make necessary. After district laws had been for some time in operation in all these new mineral regions, Territorial laws were passed, by legislatures, a majority of whose members were interested in mines, and the force and legality of the local codes were sustained, their more important features adopted and their usefulness much extended. The laws of Virginia City District, Nevada, including the famous Comstock ledge, were adopted at a miners' meeting on September 14th, 1859. "Two hundred feet on the lead," was the size allowed for a quartz claim, and each such claim must receive a name, and be properly recorded within ten days, after which three days of work each month were required to hold it. "Hill and surface claims" might be one hundred feet square, and "ravine and gulch claims" were "not more than one hundred feet wide," and extending "from bank to bank." The laws were quite long and minute, and were to be posted in two conspicuous places in the district, and another copy must be kept in the Recorder's office, subject, as was his "book of claim entry," to inspection whenever desired.

Reese River Mining District, Nevada, had a code adopted in 1864, at which time the district was nearly square, and twenty miles in extent from south to north. Fifty claim owners by signing a written notice of intention, could call a meeting of all the miners to depose the Recorder. That officer had the right to appoint a number of deputies, the district being so large. A special meeting to discuss amendments or changes in the organic law of the district might be called at any time by a written application signed by twenty-five miners, and fixing a time of meeting at a sufficient interval to permit general attendance. This rule differed from the California custom in many districts, which custom required those who for any reason desired a change in the laws to summon personally all the miners in the place, so that they might attend if they chose. The full code of this district will be found in the Appendix.

The Territorial Legislature of Idaho, December 26, 1864, adopted a mining code, chiefly made up from previous local enactments in the districts that had been organized, chiefly by miners from California, but to some extent by persons from Colorado and Utah. One of its features was a "fine of not more than $1,000, or imprisonment of not more than ninety days, or both," for destroying claim notices or removing the boundaries to a miner's plot.

The Arizona Statute of 1865 contains 53 sections, some of them very long, and is drawn up with much legal skill and mining experience. It shows to a greater extent than in California the survival of Mexican usages. Both Arizona and New Mexico exhibit marked traces of this influence. The statute of 1865 confirms and legalizes the previous acts of local districts. It uses the Spanish term "denunciation" of a claim, for the act of "taking it up," or "location," in the American phrase. It provides for placer mines of various sorts, for dry diggings and for quartz. It orders that the Recorder, when he examines a new claim for the purpose of describing and registering it, shall be permitted to take three specimens from different parts of the lode, and shall forward the same to the Territorial capital, to be kept until they are needed for the mineralogical cabinet "of the future University of Arizona." This is a very interesting provision, not found in any of the enactments of other Territories and States, and it would be pleasant to know that the plan had been carried out in Arizona.

The Oregon Statute, of about the same time, recognized all the local laws "in relation to mining rights, placer claims and town lots in mining camps;" it declared ditches and mining flumes to be real estate and mortgages of interest in placer mines to be "mortgages of chattels." The county clerk of each county was declared *ex officio* recorder of the mines, and ordered to appoint deputies in the several districts.

Montana, partly under the influence of Idaho enactments, partly under influences from the Mississippi valley, organized multitudes of thriving camps in the great Helena gold basin. The years 1863, '64, and '65 were the palmy days of Montana placers. Vigilantes were forced to organize in order to save society from being trampled under foot by one of the most remarkable bands of desperadoes that the Far West has ever known. No "dime novel" ever contained so thrilling chapters as the plain narrative of the destruction of "Plummer's Road Agents" by silent and secret Vigilantes, in the dead of the winter of 1863--4. The land laws of "Alder Gulch," one of the most famous of Montana Districts, were adopted at a mass-meeting of the miners shortly after the work of the Vigilantes had been accomplished, and protection to life and property secured. These laws appear in the Appendix to serve as an example of the Rocky Mountain camp laws, even as those of Columbia District, California, serve to illustrate the Sierra Camps.

We can still further understand the contradictory regulations of different Districts and Territories by observing how greatly the size of the "location" varied in the year 1866, when Congress began to legislate in regard to these mineral lands. By examination of the various local codes, and the statutes, acts, etc., of the States and Territories wherein mining had become a great industry, we find that in 1866 the size of the location allowed for quartz claims in various places named was as follows: Copperopolis District, California, 150 feet by 250 feet on each side of the vein; Ophir District, California, 100 feet by 100 feet on each side; Hardscrabble District, California, "fifty feet on the ledge for each location," amended to 200 feet soon after; Sierra County, 250 ft. x 250 ft.; Tuolumne County, 150 ft. x 150 ft.; Nevada County, 100 ft. x 200 ft. in total width; Nevada State, 200 ft. x 200 ft.; Idaho, 200 ft. x 100 ft.; Oregon 300 ft. x 150 ft.; Arizona, 600 feet square. Angel's District, California, had the "square," or Spanish location system, not allowing the owner to follow the "dips, spurs and angles" of his vein of ore beyond the limits of his claim. The usual Spanish law allowed in 1866 an entry of 200 varas, or 550 feet, each way. The San Antonio District, Nevada, provided that any quartz claim opened to a depth of fifteen feet "forever remains the property of the claimant," and "can never be re-located." Laws of this sort, passed by a few districts, operated injuriously to the progress of the places, driving away numbers of workers and much capital. In addition to these requirements, Nevada required, in 1866, the excavation of 50 cubic feet of rock for each 200 feet of claim, each year. Oregon required $50 in work each year on each claim. Idaho made the work tax $100, which secured perpetual ownership. Arizona demanded a thirty-feet shaft, or a fifty-feet tunnel, to hold a claim for two years. Thirty days of work each year was needed after the expiration of that time. The minor regulations in all these State and Territorial Acts show the respect still paid to District customs and laws. Whenever placers were found, the regulations thereof at once assumed the general form of the early camp codes that we have been considering. As the specimens of laws printed in the Appendix to this pamphlet will sufficiently show, the changes that occurred were all in the line of larger claims and the demands of capitalists and associations. The confinement of a prospector to one claim in a District fell slowly into disuse, as the prospector became more and more a man who "took up claims" for others as well as for himself.

In 1866, Congressional legislation sanctioned these local laws, whenever "not in conflict with the laws of the United States." The legislation of Congress in subsequent years has continued in the same general lines, has specified with increasing definiteness the accepted modes of locating and holding mining claims, and the manner of obtaining a U. S. patent to mineral lands, always recognizing, however, the essential features of those laws framed in the early California camps.

The permanent influence of the miners' civil code is further manifested in every portion of the mining law of the United States to-day. The statute of May 10th, 1872, and the "Revised Statutes" assert that mineral lands of the government are "open to exploration, occupation and purchase," under regulations "prescribed by law and according to the local customs of miners in the several mining districts." "Mining claims," it provides, "upon being in quartz or other rock," shall be governed "by the customs, laws and regulations in force at the date of their location." Previous locators are protected, and only ten dollars worth of work per year on each claim is required, whereas on locations made after the passage of the Act one hundred dollars worth of work is essential. But the influence of local laws is evident throughout. No government title to mineral land can be obtained until the claimant proves "a compliance with the mining rules, regulations and customs of the mining district, state or territory in which the claim lies." When his papers, mining records, certificates of location, claim notices, local recorder's certificate, etc., are accepted by the government land officer, the claimant is allowed to purchase the tract at five dollars per acre for vein or lode claims, and half that sum for placer claims.

Thus, for many years, the process of establishing new camps, and the process of determining by judicial acts and decisions, in courts, legislatures and congress the relative value of those local laws, went on simultaneously, ever evolving from crudeness and seeming contradiction the higher forms of mining jurisprudence. Everywhere life and energy, working on a gigantic scale, have plowed furrows into the institutional "bed rock" of Western society; the placer-miner's rude "rocker," hewn from the great sugar-pines of the Sierra, the tall firs of the Cascades, the twisted spruces of the Black Hills, the snow-weighted cedars of Kootenay, have cradled infant Territories and young, lusty States.

Leaving this past and coming to the actual camp life of the present year, we discover that the region of the United States where the spirit of the earlier system is strongest is that vast mountain realm which extends north of Nevada, Utah and Colorado.

> "In that desolate land and lone
> Where the Big Horn and Yellowstone
> Roar down their mountain path,"

where it is a wilderness of crags, narrow vales, and high plateaus for hundreds of miles west from the place that saw the heroic death of "the white chief with yellow hair," the "mining-camp" is the beginning of government. To-day, these camps cluster most thickly about that enormous mass of rugged ranges from which the Columbia, Peace, Fraser, Toiyepe, Marias, Madison, Clearwater, Saskatchewan, and dozens of other rivers flow, towards all points of the compass.

Here is probably the last hope for discoveries of wealth-bearing placers of any great extent, and here, last winter and spring, began that exciting "rush" to the Coeur d'Alene region, of which, after a summer of conflicting reports, the result appears to be somewhat disastrous. The thousands that went thither, however, found gold in many places, and organized temporary camps. One on Pritchard and Eagle Creeks, Shoshone County, adopted its local code concerning claims during the first week in March of the present year, 1884. The greatest of changes manifest from the common rules of elder camps was in the size of the claims. All locations on lodes of quartz, made so as to conform with the U. S. mining law of 1872, are to be 1,500 feet in length and 600 feet in width. All placer miners are allowed to locate twenty acres, so situated that neither dimension of the tract shall exceed eighty rods. The friendly and socialistic miners of the California camps of the "Flush Period" have little part in this later development; the capitalist has come to the front, and desires the whole of a quartz lode for his mill, or sufficient placer ground to justify the use of hydraulic methods. This is further shown by section third, which provided that authorized agents for capitalists may locate and record claims for them. Such a thing was never heard of in the early mining days on the Pacific coast, but the professional prospector and locator of claims for others is a prominent figure in western camps. "Give me a grub stake an' I'll locate ye a dozen good mines," is the appeal made to each "tender-foot" as a stranger is affectionately termed. Still another section of this Coeur d'Alene code allows men to take up one claim in each of the separate gulches where mineral is found.

In respect to claim work and assessment, another series of changes from the camps of twenty years ago is manifest. One hundred dollars worth of work must be done within the first year after location, and twenty dollars worth of work for each month between June and November in subsequent years. Road making and cabin building are allowed to count on the assessment, at the rate of five dollars per day. A claim must be recorded within fifteen days after location, the district being so large that a shorter time-allowance would work hardship in many cases. The laws announced in regard to riparian rights and privileges are much the same as those of thirty years ago. Miners may unite their claims for the purpose of more convenient working. This, also, is a common expedient. Most interesting of all, we observe that the system of arbitration still prevails. "Each disputant is to be allowed an equal number of arbitrators, and in case of a tie on the decision each arbitrator shall have power to call in an assistant." Twelve miners by ten days' written notice can call a meeting to change the laws of the District. The officers are "chairman" and "claim-recorder." This interesting code appears in the Appendix, and can be compared with those of earlier camps.

The laws of placer and quartz districts in the Saw Tooth Range, along the Salmon, and in the Great Bend of the Columbia, districts actually organized within the past year, and in existence at the present time, might easily be added to those of the Coeur d' Alene miners, but they present few variations from the types already illustrated at sufficient length. Some of the more general aspects of the subject, however, seem to deserve attention. The outward spread of the camps, from their starting point in California, has been spoken of, and also the crystallization of crude local laws into more general enactments, which process the quotations in the Appendix, from decisions of the State Courts and from legislative acts will more clearly show. But the living force and influence of these local institutions thus created needs further emphasis and illustration, even at the risk of appearing to be "re-working the placer."

Over a period of thirty-six years of ardent and picturesque life the story of these small frontier settlements, called "Mining Districts," can be said to extend. We have seen to some degree the nature of their enactments, based upon the facts that the government allowed entry upon the public mineral lands of the United States, and that until 1866 nothing was done by Congress to limit or define that right. Thus, for almost twenty years after the discovery of gold in the California gulches, the seekers for the precious metal were therefore left to their own resources in the matter of district laws. This was a creative opportunity such as the miners of no other race had possessed in modern times. Australian miners found that government officials and tax collectors, assize courts and the machinery of civilized society were at Ballarat, Bendigo and Mount Hope before the ground was fairly prospected. An efficient and satisfactory system was thus extended over the mines, but it was far from being the creation of the miners themselves. In California, on the contrary, so universal was the habit of self-organization, so generally acknowledged were the benefits of the system, that often the first thing that the dwellers at the county seat knew of a gold discovery was the news that a "new district had been formed" and given a name, and that twenty or fifty claims, as the case might be, had been "taken up" the first day.

The curious and typical western character, not as yet quite understood in literature, and sadly caricatured in most cases, is not the claim owner, not the red-shirted miner in the camp—swinging his twelve-pound pick of glittering steel, four feet from tip to tip, the weapon for a Titan. Grand figure, indeed, is the jovial and "well-located" miner, but the true romance of the Far West lingers more with the miners whose claim is exhausted, or has failed to "pan out;" it lingers most of all with the wandering *prospector*, the pioneer of Western civilization, who explores gulch after gulch, mountain range after mountain range, and toils across deserts, and poles his canoe up swift torrents, and rides his "*Cayuse*" into regions where Indians and grizzlies dwell, too often leaving his bones to keep guard in the wilderness.[1] But this typical "prospector," wedded to his free-mining customs, carries with him wherever he wanders the spirit of a law-abiding American citizen; and to-day, as of old, the first thought, when he finds a rich gulch and companions gather about him, is "camp-organization."

Another fact clearly evident to students of the institutional development of the Far West is the unusual permanence of the forms thus created. In more than three thousand Districts, many of them since grown into thriving cities, land

[1] "Few know or have ever paused to think how many men have perished in these wild crusades searching after gold in remote localities. Embarking in frail crafts, some have suffered shipwreck on their way to or back from Gold Bluff; some, overcome with exposure and fatigue, lay down and died in the gloomy woods of British Columbia. Some were swept to death over the rapids or engulfed in the treacherous eddies of the Fraser. Of the more adventurous, not a few battled with Indians on the distant frontiers, while others, vainly struggling, sank under their burdens in the snow of the Sierra or perished from thirst far out on the deserts, where their uncoffined bones lie bleaching to this day." Editorial in "Mining and Scientific Press," San Francisco, January, 1884.

laws of the nature we have been describing lie at the foundation of their polity. There were not, at any one time, more than a thousand camps in existence, but where some perished others took their places, while the living link of camp with camp, district with district, code of the California of 1850 with code of the Montana of 1864, or of the Idaho of 1884, was none other than the wandering prospector, who had helped to make the laws of many such settlements.[1] Although the material at hand for a study of the mining camp land laws is sufficient to reveal their nature and relationship, yet the student of the period is forced at last to vain regret over the loss of the records of so many famous camps, the obscurity of the recollections of so many pioneers. Men who are making history, who are founding States, are not apt to display any zeal in treasuring up the documents which explain their deeds. If a camp were well-governed, what mattered it to the "Argonauts" whether a copy of their proceedings were preserved? Changes, removals, decay of camps, fires in mountain-towns, and a host of disasters have overtaken the crude and plain enactments of hundreds of communities. In general terms it may be said that the "rules and regulations" of nearly a hundred camps on the Pacific Coast or in the Rocky Mountain region are still preserved in print or in manuscript, and that fragments of evidence concerning the usages of many other camps yet remain, buried in law

---

[1] It is impossible to describe, in the space at our disposal, the great "mining excitements" that scattered broadcast the local laws of the early Districts. The following, however, were some of the most famous ones: Gold Lake, "the first stampede," summer of 1850; Gold Bluff, spring of 1851; Kern River, '54–'55, which took 5,000 miners to a region where most of them failed to pay expenses. Fraser River in 1858, took 18,000 men from California, and San Francisco real estate lost from 25 per cent to 75 per cent. The terrible hardships of these bold explorers did not lessen the force of the Washoe excitement of two years later, when 12 mines on the Comstock paid dividends before 1865, and 2,988 mines did not. In 1869 came the White Pine rush, then the Bodie excitement, then Snake River, and others almost yearly till the last, that of the Cœur d'Alene, near the borders of British America.

reports and official documents. But there can be little doubt concerning the real sufficiency of this material. The general nature of camp law appears from the evidence accumulated in the course of this "study," hardly less than if that evidence consisted of tabulated reports from all the camps of the period. In itself the material which can be accumulated is ample, and only when the mass of what has perished is taken into consideration does it seem in anywise inadequate. For further evidence concerning the practical workings of the camp codes in relation to claims, and their use or forfeiture, we must examine the reports of travelers and take the testimony of the pioneers themselves. There is hardly an exception to the opinion that law and order prevailed, that "landed property" on the mining-camp basis was protected, that the attempt to make use the only title-deed was successful in all the early placers.

But we cannot pursue our subject into these broader fields. The social and intellectual aspects of the Mining Camp, and its relationship to the growth of organized society in the region where for a time it ruled paramount, belong to the book-maker, not to the pamphleteer, for this higher problem is complicated with elements contributed by alien races and different civilizations. The Camp, the Mining District, the commonwealth of freemen settled for a time in close companionship under the lofty snow peaks, breaking each other's bread, and sharing each other's blankets—*this* must be accepted as a potent factor in all the beginnings of social order, over an extent of territory five times the size of France. The roots of its growth lie deep-tangled in the soil of *Lex Saxonum*, and capitularies of Karl the Great; they spring more nearly from New England town-meetings and parish-meetings of the South, and settlers' associations of the West, but, unlike the latter, their influence has outlasted the conditions which gave them birth.

Of the land laws of these tent-dwellers on Mormon Island, these cabin dwellers of Garrote, Cat Camp, Boneyard, Moccasin Creek, Big Humbug, Grizzly Scare, and later California

Districts bearing even more startling names, the least that history will say is that they were able to win recognition in courts of higher jurisprudence, and formed a controlling factor in the creation of American mining law. The Law Reports of California, Arizona, Nevada, Oregon, Colorado, Montana, Utah, Wyoming, and Idaho bear evidence to this. There is hardly a mining case in all the volumes of these Reports that does not depend more or less upon the "land laws" of the district in which the disputed claims are situated. Rights over a plot of definite size, located according to local law, "staked" or otherwise marked, "registered," held by "work assessments" and never "abandoned;" rights over all running water not otherwise appropriated; rights over quartz ledges, even on the same ground before occupied by placer claimants; superior rights of the mineral seeker over the farmer or herdsman upon public lands—these are some of the "points" of local law that later courts from time to time enforced. A noteworthy example is in the case of the St. Louis Smelting and Refining Co. *vs.* Kemp & Nuttall, on appeal from the U. S. Circuit Court of Colorado, in which the U. S. Supreme Court this year rendered a decision of importance. A "Location" is held to be that quantity of mining ground which one person may legally acquire by location, in one body; a "claim" may embrace a dozen such locations acquired by purchase, provided they are contiguous, and the required annual expenditure may be upon any portion of the claim, or be at a distance from the claim itself, as when the labor is performed for the turning of a stream, or the construction of a flume.[1] In the noted case of Sparrow *vs.* Strong before the Supreme Court in 1865, Chief Justice Chase had said: "We know that the territorial legislature has recognized by statute the validity and binding force of the 'rules, regulations, and customs of the mining districts,' and we cannot shut our eyes to the public history which in-

[1] Decision given by Mr. Justice Field. Compare section 2,330 of the Revised Statutes of the United States.

forms us that under this legislation, and not only without interference from the general government, but under its implied sanction, vast mining interests have grown up." The decisions of the present time show the same respect for vested rights, although the passage in 1866 and 1872 of general laws, relative to the purchase or pre-emption of mineral lands of the United States, has greatly lessened the scope of "district law influence."

Nevertheless, the permanent place which such local law occupies is shown in the fact that in many cases at the present time men from a district agree to abide by the United States mineral land laws, make a few minor land regulations, and keep up the district organization for other purposes. Prospectors and temporary holders of claims will always need local enactments to prevent quarrels, and these enactments they will continue to ordain in "miners' meetings" for many years to come. Further than this, the permanent influence of camp laws is clearly manifest in the organic life of such typical California mountain towns as El Dorado, once Hangtown; Nevada City, once Caldwell's Upper Store; Shasta City, once Redding Springs; Downieville, once a group of tents; Sonora, once a half-Mexican village. Each one of these places is now a county seat, and some are towns of several thousand inhabitants, but in most cases their incorporation as towns was done by the "miners of the district," while the mining industry was yet predominant, and their organization of to-day is more simple, more direct and more dependent upon popular will than are most municipal systems.

So far as California is concerned, the truth remains that, long after the State was divided into counties and townships, the camps, whose usages we have discussed, were flourishing undisturbed under their local laws, and with their local recorders, or other presiding officers. Social, political, and literary elements of primal importance the study of this system reveals. The instinct of the novelist and poet has already seized hold of a few of the more effective and picturesque features of the early camp life and district law, but a broad realm for true word-artists is as yet unconquered. As a chapter in political science the place of the full story of these districts will be recognized when the right men, trained in schools of comparative institutional history, come to the writing of the growth and development of communities west of the Mississippi. The student of sociology will say, as he investigates the organization of these early camps: "Here are glimpses of Jean Jacques Rousseau, and the 'Social Contract' theory; here is a harmless and altogether new form of socialism; here, for a brief space, all the world was 'lawless' according to strict legal interpretations, but wonderfully blessed with the essence and spirit of true self-government." Because more than a hundred thousand young, able-bodied American citizens bent their backs to the miner's mighty toil, and for years "went camping," under skies bluer than Italy's, in the torrent-watered cañons of mountains loftier than the Alps; because of their brawny strength, their splendid vitality, their terrible earnestness, the laws they formulated in "miners'-meetings" held under no tent roof, but in open air, like the "*Guirimears*" of ancient Cornwall, were laws that have an abiding historical significance for all Americans.

APPENDIX.

Examples of Mining District Laws,

WITH QUOTATIONS FROM TERRITORIAL AND STATE ENACTMENTS, AND FROM REPORTS OF IMPORTANT TRIALS, SHOWING THE RESPECT PAID TO DISTRICT LEGISLATION.

PLACER LAWS OF COLUMBIA DISTRICT, TUOLUMNE CO., CALIFORNIA, 1854.

Article I. The Columbia Mining District shall hereafter be considered to contain all the territory embraced within the following bounds: Beginning at the site of McKenny's old store on Springfield Flat and running in a direct line to a spring on a gulch known as Spring Gulch—said gulch running in a southern direction from Santiago Hill. Thence in a direct line to the angle of the road leading from Saw Mill Flat to Kelly's Ranch, near Woods Creek. Thence along the ridge on the west of Woods Creek to the southern bounds of Yankee Hill District. Thence, following the ridge to the high flume between Columbia and Yankee Hill. Thence, following the New Water Company's ditch to Summit Pass. Thence in a direct line to the head of Experimental Gulch, including said gulch. Thence, following the upland to the head of Fox Gulch, and including said gulch. Thence, following the upland around the head of Dead Man's Gulch to the site of the Lannesdale saw mill. Thence in a direct line to the place of beginning.

Art. II. A full claim for mining purposes on the flats or hills in this district shall consist of an area equal to that of one hundred feet square. A full claim on ravines shall consist of one hundred feet running on the ravine, and of a width at the discretion of the claimant, provided it does not exceed one hundred feet.

Art. III. No person or persons shall be allowed to hold more than one full claim within the bounds of this District, by location; nor shall it consist of more than two parcels of ground, the sum of the area of which shall not exceed one full claim; *Provided* that nothing in this article shall be so construed as to prevent miners from associating in companies to carry on mining operations, such companies holding no more than one claim to each member.

Art. IV. A claim may be held for five days after water can be procured at the usual rates, by distinctly marking its bounds by ditches, or by the erection of good and sufficient stakes at each corner, with a notice at each end of the claim, followed by the names of the claimants, and by recording the same according to the provisions of article ten.

Art. V. When a party has already commenced operations upon a claim, and is obliged to discontinue for want of water, or by sickness or unavoidable accident, the presence upon the ground of the tom and sluices, or such machines as are employed in working the claim, shall be considered as sufficient evidence that the ground is not abandoned, and shall serve instead of other notice; the bounds of the claim being still defined.

Art. VI. Claims shall be forfeited when parties holding them have neglected to fulfil the requirements of the preceding articles, or have neglected working them for five days after water was procurable at usual rates, unless prevented by sickness or unavoidable accident, or unless the miners have provided by law to the contrary. [This peculiar clause may be construed to mean that if the miners thought the owners of "water-privileges" were charging exorbitant rates, they could temporarily enact that cessation of work should not cause forfeiture.]

Art. VII. Earth thrown up for the purpose of working shall not be held distinct from the claim from which it was taken, but shall constitute part and parcel of such claim.

Art. VIII. Water flowing naturally through gold-bearing ravines shall not be diverted from its natural course without the consent of parties working on such ravines, and when so diverted it shall be held subject to a requisition of the parties interested.

Art. IX. No Asiatics shall be allowed to mine in this district.

Art. X. Any or all claims now located or that may hereafter be located and worked, can be laid over at any time, for any length of time not to exceed six months, by the person or persons holding the same appearing before the recorder of the district, with two or more disinterested miners, who shall certify over their own signatures that the said claim or claims cannot be worked to advantage, and by having the same recorded according to the laws of the District; and by paying a fee of one dollar; *Provided*, that each claimant shall sign the record in person or by a legal representative, stating at the time that said claim is held by location or by purchase.

Art. XI. There shall be a Recorder elected who shall hold his office for one year from the date of his election, or until his successor be elected, whose duty it shall be to keep a record of all miners' meetings held in the district; to record all claims; and to call miners' meetings by posting notices throughout the district whenever fifteen or more miners shall present him with a petition stating the object of the meeting: *Provided*, that in the absence of the Recorder the above-named number of miners shall not be disqualified to call such a meeting. He shall at all proper times keep his record book open for inspection.

Art. XII. No company or companies of miners who may occupy the natural channel through any gulch or ravine for a tail-race or flume shall have the exclusive right to such a channel, to the exclusion of any company of miners who may wish to run their tailings into the same.

Art. XIII. Any party or parties locating claims in gulches, where such flumes or tail-races exist, shall first confer with the owners of the same, . . . and in case of disagreement each party shall choose two disinterested miners and the fourth shall choose a fifth, who may determine the matter or matters in dispute.

Art. XIV. Any company or companies of miners shall have the right to run their water and tailings across the claim or claims below them, if it can be done without injury to the lower claims.

Art. XV. The limits of this District shall not be altered without the consent of a regularly called mass-meeting of the miners of the District.

Art. XVI. No miners' meeting held outside of [the town of] Columbia for the purpose of making laws to govern any portion of the district, or to amend these laws in any manner, shall be considered as legal.

Art. XVII. All mining laws of this district made previous to the foregoing are hereby repealed. [Passed in 1854, and in full force in 1866].

PLACER LAWS OF ALDER GULCH, MONTANA, 1864.

PREAMBLE.

Whereas, the laws now in force in Fairweather District, Madison County, Montana, have proved insufficient to protect the rights of the miners of said District;

And whereas the rights and interests of the miners of the District are of such a nature as not to admit of a resort to the tedious remedy of the ordinary process of law for every violation of those rights;

Now, therefore, we, the miners of said District in public meeting assembled, in pursuance of legal notices, for the purpose of defending our rights and duties and the protection of our several interests, do hereby resolve and declare that the rules and provisions following shall be the law of Fairweather District from this date of enactment, viz: September 16th, 1864.

ARTICLE A.

Section 1. The officers shall consist of a President and Secretary who shall hold their offices for the term of six months, or until their successors are duly elected and enter upon the discharge of the duties of their office.

Sec. 2. It shall be the duty of the President to call a meeting of the miners of the District at any time on the written application of five claim holders of the district, of which he shall give three days' notice . . . by written or printed advertisements . . . posted at three of the most public places in the District, and he shall preside over the meeting.

Sec. 3. It shall be the duty of the Secretary to attend all meetings . . . keep a true record . . . and file the same with County Recorder.

Sec. 4. After suit commenced in any case wherein the title to a claim is called into question, neither party shall be held liable to represent said claim during the pendency of litigation, but the same shall be deemed to be represented in favor of the real owner by operation of law.

[*Sections* 5, 6, 7 and 8 refer to claims held by purchase, by lease or by pre-emption, authorizing all three forms when properly recorded.]

Sec. 9. Every claim shall be considered as pre-empted upon which the pre-emptor, a purchaser, shall, by himself, his agent, or hired hands, perform three full days' work in each week . . . ; *Provided*, that each and all of such claims be recorded.

[*Sections* 10 and 11 define and limit the rights of co-partners, in claims and ditch enterprises.]

Sec. 12. The absence of any person from the district shall not impair or invalidate his rights: *Provided*, his interests are represented by his partners, agents, or men in his employ.

Sec. 13. The rights of a sick member shall be respected during his illness, and the certificate of a physician shall be sufficient evidence of such illness.

Sec. 14. Any miner who shall have expended $600 on his claim, or who, for want of money for opening the same, is unable to represent it according to law, shall have the privilege of working on any other claim in the district in order to raise money to enable him to fully open up his own claim: *Provided*, he shall put up notices on his own claim stating where he is at work, and his rights shall be respected during the time he is so at work for others.

Sec. 15. It shall and may be lawful for any person or company to dig a drain ditch through the claim of any person or persons and . . . [they] shall have a lien upon any and all such claims thoroughly drained for a just and equal proportion of the cost thereof . . .

Sec. 16. The water in any creek or gulch shall belong exclusively to the miners of that creek or gulch.

Sec. 17. Each gulch claim shall be entitled to one sluice-head of water, of not less than twenty inches . . . and such additional quantity as may be necessary . . . if not used to the injury of others.

Sec. 18. The interest of the holder or holders of any creek or gulch claim is hereby declared to be a chattel interest, consisting of the right to the possession of the land and water thereupon inseparable and indivisible, except by the consent of the party or parties in interest made in due form of law, and then only to such an extent as shall not impair nor infringe upon the rights of others.

Sec. 19. No person or persons in company shall have the right . . . to claim and hold an exclusive right or privilege in or to any portion of the water in any creek or gulch in the district . . . and any ditch, pipe, channel, flume or other means of conveyance . . . by which water may be directed from its original channel without leaving in creek or gulch the quantity of water belonging to each claim, is hereby declared a public nuisance, and may be abated immediately . . . in accordance with the laws of this Territory.

[*Sections* 20, 21, 22 and 23, all refer to unlawful obstructions, tailings, leakage from flumes and protection of drain ditches. The remaining four sections of this Article ordain that claims not earned according to the district customs shall be forfeited on November 1st, of each year, and liable to pre-emption by any person at any time before May 1st, when work in the district was commenced; they repeal all laws, parts of laws, and rules, customs, and regulations, in conflict with the present code, and they ordain that the new laws shall go into immediate effect.]

ARTICLE B.

Section 1. Bar mining claims shall consist of 100 feet up and down the gulch or creek, and running back the width of the bar.

Sec. 2. Creek claims shall be 100 feet in length, and including the bar, or creek bottom, and the head of the stream.

Sec. 3. All discovery claims shall be safely held whether worked or not.

Sec. 4. The centre of the creek shall be the line.

QUARTZ LAWS OF REESE RIVER DISTRICT, NEVADA TERRITORY, 1864.

Section 1. The district shall be known as the Reese River Mining District, and shall be bounded as follows, to wit: On the north by a distance of ten miles from the Overland Telegraph Line, on the east by Dry Creek, on the south by a distance of ten miles from the Overland Telegraph Line, and on the west by Edward's Creek, where not conflicting with any new districts formed to date.

Sec. 2. There shall be a Mining Recorder elected on the first day of June next, for this District, who shall hold office for one year from the seventeenth of July next, unless sooner removed by a new election, which can only be done by a written call, signed by at least fifty claim holders, giving notice of a new election to be held after said notice shall have been posted and published for at least twenty days in some newspaper published in or nearest this District; and the Recorder shall be a resident of this District.

Sec. 3. It shall be the duty of the Recorder to keep in a suitable book or books, a full and truthful record of the proceedings of all public meetings; to place on record all claims brought to him for that purpose, when such claim shall not interfere with or affect the rights and interests of prior locators, recording the same in the order of their date, for which service he shall receive one dollar ($1) for each claim recorded. It shall also be the duty of the Recorder to keep his books open at all times to the inspection of the public; he shall also have the power to appoint a deputy to act in his stead, for whose official acts he shall be held responsible. It shall also be the duty of the Recorder to deliver to his successor in office all books, records, papers, etc., belonging to or pertaining to his office.

Sec. 4. All examinations of the record must be made in the full presence of the Recorder or his deputy.

Sec. 5. Notice of a claim of location of mining ground by any individual, or by a company, on file in the Recorder's office, shall be deemed equivalent to a record of the same.

Sec. 6. Each claimant shall be entitled to hold by location two hundred feet on any lead in the district, with all the dips, spurs, and angles, offshoots, outcrops, depths, widths, variations, and all the mineral and other valuables therein contained—the discoverer of and locator of a new lead being entitled to one claim extra for discovery.

Sec. 7. The locator of any lead, lode or ledge in the district shall be entitled to hold on each side of the lead, lode or ledge located by him or them, one hundred feet; but this shall not be construed to mean any distinct or parallel ledge within two hundred feet other than the one originally located.

Sec. 8. All locations shall be made by a written notice posted upon the ground, and boundaries described, and all claimants' names posted on the notice.

Sec. 9. Work done on any tunnel, cut, shaft or drift, in good faith, shall be considered as being done upon the claim owned by such person or company.

Sec. 10. Every claim (whether by individual or company) located, shall be recorded within ten days after the date of location.

Sec. 11. All miners locating a mining claim in this district, shall place and maintain thereon a good and substantial monument or stake, with a notice thereon of the name of the claim, the names of the locators, date of location, record and extent of claim. It is hereby requested that owners in claims already located do comply with the requirements of this section.

Sec. 12. The Recorder shall go upon the ground with any and all parties desiring to locate claims, and shall be entitled to receive for such service one dollar for each and every name in a location of two hundred feet each.

Sec. 13. It is hereby made the duty of the Mining Recorder upon the written application of twenty-five miners, to call a meeting of the miners of the district by giving a notice of twenty days through some newspaper published in the Reese River District, which notice shall state the object of the meeting, the place and time of holding the same.

Sec. 14. The laws of this district passed July 17th, 1862, are hereby repealed.

Sec. 15. These laws shall take effect on and after June 4th, 1864.

PLACER LAWS OF CŒUR D'ALENE DISTRICT, IDAHO, 1884.

The local laws of the Cœur d'Alene Mining District located on Pritchard and Eagle Creeks, Shoshone County, Idaho Territory, adopted by the first permanent miners in that region, were as follows :

Section 1. All locations on lodes of veins of quartz to conform with the United States laws of May 10, 1872, as nearly as practicable, viz : Fifteen hundred feet (1500) in length, by six hundred feet in width.

Sec. 2. Placer mining claimants shall be allowed twenty acres, to be located so that the length shall not exceed 80 rods.

Sec. 3. Each location shall be represented by the locator or his authorized agent in locating and recording.

Sec. 4. No person shall be restricted to one claim, but may locate one claim in any stream or gulch where vacant ground may be found. But no person shall be allowed to locate more than one claim on the same stream or gulch. Persons shall not be prohibited from holding claims acquired by purchase.

Sec. 5. Claimants shall have one year from the first of January succeeding the date of location to work their first annual assessment, which shall be one hundred dollars. Each year thereafter claims shall be represented by twenty dollars' worth of labor each month after the first of June until the first of November after the first year's assessment. Furthermore, all claims shall be considered laid over from the first of November to the first of June. All necessary work, such as making roads or trails, building houses, or any improvements in opening or working a claim, will be allowed five dollars per day as assessment labor.

Sec. 6. Claimants will be required to record their claims in the district record within fifteen days from the date of location.

Sec. 7. The oldest or first claimants shall have the first privilege of water, but shall not prohibit others from using the surplus water, and all claimants shall be required to return the water to the channel of the stream for the benefit of those below.

Sec. 8. Several miners may form a company for the purpose of opening and working mines in the placer claims, when such are contiguous, and the labor performed by said company shall represent their several claims, although the amount of labor for the convenience of opening and working may be done on one claim.

Sec. 9. Difficulties arising between parties in the mining district shall be settled by arbitration, each disputant to be allowed an equal number of arbitrators, and in case of a tie on decision said arbitrators will have power to call an assistant.

Sec. 10. All claims located prior to the date of the adoption of these by-laws shall be respected just the same as those made after said date.

Sec. 11. The records of the Cœur d'Alene Mining District, in Shoshone county, Idaho Territory, shall be kept at the house of A. J. Pritchard, Recorder, near the confluence of the Eagle and Pritchard Creeks.

Sec. 12. On the written application of twelve or more miners, the Chairman shall cause three notices to be posted up in three conspicuous places, giving ten days' notice of a meeting, said notices to specify the object and business to be transacted at such meeting. To make any changes in the present by-laws between the 1st of November and the 1st of June the following year shall be illegal.

Sec. 13. These laws shall take effect from this date, and any laws or regulations previously enacted that conflict with these laws shall be considered repealed.

Judicial Decisions upon Local Rules, Usages and Customs.

"The Code permits evidence of the customs established in mining claims, which implies a permission on the part of the State to the miner to seek whatever he choose in the mines for the precious metals, and extends to him whatever right the State might have to the mineral when found."—*Case of McClintock vs. Bryden*, 5 *Cala. Reports*, *p.* 100.

"The custom of miners is entitled to great if not controlling weight." —*Brown vs.* "49 *and* 56" *Quartz Mining Co.*, 15 *Cala. Rep.* 160.

"The quantity of ground a miner can claim by location or prior appropriation for mining purposes may be limited by the mining rules of the district."—*Prosser vs. Parks*, 18 *Cala.*, 47.

"The fact that mining laws and regulations were passed on a different day from that advertised for a meeting of miners does not invalidate them. Courts will not inquire into the regularity of the modes by which these local legislatures or primary assemblages act. They must be the judges of their own proceedings. It is sufficient that the miners agree —whether in public meeting or after due notice—upon their local laws, and that these are recognized as the rules of the vicinage, unless fraud be shown, or other like cause for rejecting the laws."—*Gore vs. McBrayer*, 18 *Cala.*, 582.

"A mining claim must be in some way defined as to limits before the possession of the work upon part gives possession to any more than the part so possessed or worked. But when the claim is defined, and the party enters in pursuance of mining rules and customs, the possession of part is the possession of the entire claim."—*Atwood vs. Fricot*, 17 *Cala.*, 27.

"Although mining ground . . . may be located in the absence of local regulations, yet the extent of such location is not without limit. The quantity taken must be reasonable, and whether it be so or not will be determined by the general usages and customs prevailing upon the subject. If an unreasonable quantity be included . . . the location will not be effectual."—*Table M. T. Co. vs. Stranahan*, 20 *Cala.*, 198; 21 *Cala.*, 548.

"Mining laws, when introduced as evidence are to be construed by the Court, and the question whether, by virtue of such laws a forfeiture had occurred is a question of law, and cannot be submitted to a jury."—*Fairbank vs. Woodhouse*, 6 *Cala.*, 433.

"A party having abandoned his claim will not be permitted to . . . reassert or resume his former interest to the prejudice of those who may have afterwards appropriated it."—*Davis vs. Butler*, 6 *Cala.*, 511.

"The book in which claims are recorded by resolution of the miners of a district may be admitted as evidence."—*McGarrity vs. Byington*, 12 *Cala.*, 426.

"Mining claims are held by possession, but that possession is regulated and defined by usage and by local and conventional rules, and the 'actual possession' which is applied to agricultural lands cannot be required in case of a mining claim."—*Atwood vs. Fricot*, 17 *Cala.*, 37.

"A miner is not expected to reside on his claim, nor build on it, nor cultivate it, nor enclose it. He may be in possession by himself, or his agents, or servants."—*English vs. Johnson*, 17 *Cala.*, 107.

"From an early period of our State's jurisprudence we have regarded claims to public mineral lands as titles. . ."—*Merritt vs. Judd*, 14 *Cala.*, 64.

"A writing is not necessary to vest or divest title on taking up a mining claim. The right of the miner comes from the mere appropriation of the claim made in accordance with the mining rules and customs of the vicinage." "The right to mining ground rests on possession only, and rights of this character need no conveyance other than a transfer of possession." "The State Act of April 13th, 1860, permits bills of sale without seal to pass title." "Transfer may as well be by simple agreement as by deed, the vendee taking possession." Cases of *Jackson vs. Feather River and Gibsonville Water Co.*, 14 *Cala.*, 22; of *Gore vs. McBrayer*, 18 *Cala.*, 582; of *McCarron vs. O'Connell*, 7 *Cala.*, 152; of *Watt vs. White*, 13 *Cala.*, 324.

"The public mineral lands of this State are open to every person who chooses to enter upon them for the purpose of mining. But this rule has its limitations to be fixed by the facts in each particular case. Certain rights of property, though not founded on a legal title, will be protected against the miner—such as houses, orchards, vineyards, and growing crops." "The right of the agriculturist to use and enjoy public lands must yield to the right of the miner when gold is discovered in the land." The agriculturist "has settled upon such lands subject to the rights of the miners who may proceed in good faith to extract any valuable metals that may be found in the lands so occupied by the settler." "The right to so enter, and mine carries with it the right to whatever is indispensable for the exercise of this mining privilege,—as the use of the land, and such elements of the freehold as water." "It carries with it the right to the wood and timber growing thereon."—See cases of *Smith vs. Doe*, 15 *Cala.*, 100; *Clark vs. Duval*, 15 *Cala.*, 88; *Tartar vs. Spring Creek W. & M. Co.*, 5 *Cala.*, 395; *Lentz vs. Victor*, 17 *Cala.*, 271.

"The first appropriator of water for mining purposes is entitled to have the water flow without material interruption in its natural channel."—*Bear River and Auburn Water and Mining Company vs. New York Mining Co.*, 8 *Cala.*, 333.

"Surveys, notices, stakes, and the blazing of trees, followed by work and labor, without abandonment, will in every case give title to unclaimed water on public lands over after-claimants."—*Kimball vs. Gearheart*, 12 *Cala.*, 27.

"The interest of the occupant of a mining claim is property, and is liable to taxation. The 'claim' is property...and may be taken and sold under execution."—*Cases of McKeon vs. Bisbee*, 9 *Cala.*, 137; *California vs. Moore*, 12 *Cala*, 56.

Legislative Acts.

"In all actions respecting mining claims, proof shall be admitted of the customs, usages or regulations established and in force in the Mining District embracing such claim; and such customs, usages or regulations, when not in conflict with the laws of this Territory, shall govern the decision of the action in regard to all questions of location, possession and abandonment."—*Laws of Nevada Territory, approved Nov.* 29, 1861. *Similar laws in Oregon, Arizona, Idaho, Montana and Colorado.*

"All conveyances of mining claims heretofore made by bills of sale or other instruments in writing, with or without seals, recorded or unrecorded, shall be construed in accordance with the lawful local rules, regulations and customs of the miners in the several mining districts of this Territory; and if heretofore regarded valid and binding in such districts, shall have the same force and effect between the parties thereto as *prima facie* evidence of sale, as if such conveyances had been made by deed under seal."

"The location and transfers of mining claims heretofore made shall be established and proved, in contestation before Courts, by the local rules, regulations or customs of the miners in the several mining districts of the Territory in which such locations and transfers were made." —*Laws of Nevada, Dec.* 12, 1862.

Extracts from United States Mining Law.

Be it enacted by the Senate and House of Representatives of the United States of America, in Congress assembled, That all valuable mineral deposits in land belonging to the United States, both surveyed and unsurveyed, are hereby declared to be free and open to exploration and purchase, and the lands in which they are found to occupation and purchase, by the citizens of the United States, and those who have declared their intention to become such, under regulations prescribed by law, *and according to the local customs or rules of miners, in the several mining districts,* so far as the same are applicable and not inconsistent with the laws of the United States.

Sec. 2. That mining claims upon veins or lodes of quartz or other rock in place, bearing gold, silver, cinnabar, lead, tin, copper, or other valuable deposits heretofore located, *shall be governed, as to length along the vein or lode, by the customs, regulations and laws in force at the date of their location.* A mining claim located after the passage of this act, whether located by one or more persons, may equal, but shall not exceed, one thousand five hundred feet in length along the vein or lode; but no location of a mining claim shall be made until the discovery of the vein or lode within the limits of the claim located. No claim shall extend more than three hundred feet on each side of the middle of the vein at the surface, nor shall any claim be limited, by any mining regulation, to less than twenty-five feet on each side of the middle of the vein at the surface, except where adverse rights existing at the passage of this act shall render such limitation necessary. The end lines of each claim shall be parallel to each other.

Sec. 3. That the locators of all mining locations heretofore made, or which shall hereafter be made, on any mineral vein, lode or ledge, situated on the public domain, their heirs or assigns, where no adverse claim exists at the passage of this act, so long as they comply with the laws of the United States and the State, Territorial, *and local regulations, not in conflict with said laws of the United States, governing their possessory title*, shall have the exclusive right of possession and enjoyment of all the surface included within the lines of their locations, and of all veins, lodes and ledges, throughout their entire depth, the top or apex of which lies inside of such surface lines extended downward vertically, although such veins, lodes or ledges may so far depart from a perpendicular in their course downward as to extend outside the vertical side-lines of said surface locations: *Provided*, that their right of possession to such outside parts of such veins or ledges shall be confined to such portions thereof as lie between vertical planes drawn downward as aforesaid, through the end lines of their locations, so continued in their own direction that such planes will intersect such exterior parts of said veins or ledges.

Sec. 5. That the *miners of each mining district* may make rules and regulations, not in conflict with the laws of the United States, or with the laws of the State or Territory in which the district is situated, *governing the location, manner of recording, amount of work necessary to hold possession of a mining claim*, subject to the following requirements: The location must be distinctly marked on the ground so that its boundaries can be readily traced. All records of mining claims hereafter made shall contain the name or names of the locators, the date of the location, and such a description of the claim or claims, located by reference to some natural object or permanent monument, as will identify the claim. On each claim located after the passage of this act, and until a patent shall have been issued therefor, not less than one hundred dollars' worth of labor shall be performed or improvements made during each year. On all claims located prior to the passage of this act, ten dollars' worth of labor shall be performed or improvements made each year for each one hundred feet in length along the vein, until a patent shall have been issued therefor; but where such claims are held in common, such expenditure may be made upon any one claim;

and upon a failure to comply with these conditions, the claim or mine upon which such failure occurred shall be opened to relocation in the same manner as if no location of the same had ever been made.—*Act of May 10th*, 1872.

The legal form of Location Notice is as follows:

Notice is hereby given that the undersigned, having complied with the requirements of the mining act of Congress, approved May 10, 1872, and the *local customs, laws and regulations*, has located fifteen hundred linear feet on............lode (twenty acres of placer mining ground) situated in.........mining district, ...county, ..., and described as follows: (Describe the claim accurately, by courses and distances, if possible; by legal subdivisions, if a placer claim is located on surveyed land.)

Located .... 188... ....

Recorded....... . 188... (Name of locator.)

[Record of location notices, in absence of a District Recorder, should be made with the proper Recorder of Deeds for the county wherein the claim is situated. The affidavits of at least two disinterested persons that all the requirements of the Congressional and *local laws* have been complied with, should also be recorded.]

The legal form for "proof of possession" where there is no "written title" contains the following: (U. S. Mining Laws, G. F., 1875, Form K.)

"That said mine was *located and has been possessed and worked in accordance with the customs and usages of miners in said district*, and in conformity with the rules and regulations regulating the location, holding and working of mining claims, in force and observed in the State of .... That there are no written records known to deponent existing in said mining district. That affiant is credibly informed and believes that the.........mine was located in the year 18..., and that if any record was made of such location, and of the names of the locators, the same has not been in existence for a long number of years past, and that by reason thereof the names of locators cannot now be ascertained, and no abstract of title from locators to the present owner can be made. That the possession of applicant and his predecessors in interest of said........ mine, has been actual, notorious and continuous, to the positive knowledge of deponent, since his residence in said mining district, *and that such possession has been perfected and maintained, in conformity with mining usages and customs, and has been acquiesced in and respected by the miners of the said district.*"

The legal form for a "protest and adverse claim" contains the following: (Form O.)

"That said lode (or placer claim) is in the ...... Mining District.

"That on the day of its location the premises hereinafter described were mineral lands of the public domain, and entirely vacant and unoccupied, and were not owned, held, or claimed by any person or party as mining ground or otherwise, and that while the same were so vacant and unoccupied and unclaimed, to wit: On the ...... day of ...... 18...., (name locators), each and all of them being citizens of the United States, entered upon and explored the premises, discovered and located the said ...... lode, and occupied the same as mining claims. That said locators, after the discovery of said ...... lode, drove a stake on said lode on the discovery claim, erected a monument of stone around said stake, and placed thereon a written notice of location, describing the claim so located and appropriated, giving the names of the locators and quantity taken by each, and *after doing all the acts and performing all the labor required by the laws and regulations of said ...... mining district and territory of ......*, the locators of said lode caused said notice to be filed and recorded in the proper books of record in the Recorder's office in said District on the ...... day of ......, 18....

"And affiant further says, that said locators, *in all respects, complied with every custom, rule, regulation, and requirement of the mining laws, and every rule and custom established and in force in said ...... mining district, and thereby became and were owners (except as against the paramount title of the United States) and the rightful possessors of said mining claims and premises.*

"And this affiant further says, that *said locators proved and establishea to the satisfaction of the Recorder of said ...... mining district, that they had fully complied with all the rules, customs, regulations, and requirements of the laws of said District, and thereupon the said Recorder issued to the locators of said ...... lode, certificates confirming their titles and rights to said premises.*"

[Points bearing on local law italicized in above quotations from U. S. mining law.]

Chief Authorities.

THE BOOKS, PAMPHLETS, LAW REPORTS, AND PERSONAL EVIDENCE OF PIONEERS, MOST USED IN THE FOREGOING PAGES.

CALIFORNIA.—Documents, 1850; publications of Pioneer Society, speeches, &c. Works of Avery, Bartlett, Beckwith, Benton, Brooks, Burnet, Buffum, Carson, Capron, Dunbar, Delano, Farnham, Gerstäcker, Helper, King, Marryat, Palmer, Taylor, Robinson, and other observers of the mining era. Debates in State Convention of '49, also articles on "California" and "Gold" in periodicals. (See Poole's Index.)

"DISTRICT" GOVERNMENT AND CAMP LAWS.—Codes of all the camps of Tuolumne County, and many in Nevada, Amador, Calaveras, Placer, Sierra, Shasta, and El Dorado, also in the State of Nevada, and the territories of Idaho and Montana, principally from county histories and early directories, partly from private correspondence, and to some extent from personal observation.

LAW.—U. S. Mining Laws of 1866, 1872, and subsequently; also, decisions of Secretary of the Interior. "Law of Mines and Minerals," Weeks, S. F., 1876. Land law pamphlets, arguments of Hon. Mr. Dwinelle, Horace Hawes, Judge Wilson and others in early California cases. Bainbridge's "Law of Mines." Congdon's "Mining Laws and Forms," S. F., 1864. "Legal Title to Mining Claims and Water Rights in California," by Gregory Yale, San Francisco, 1867. "Mining Laws of Mexico," by Rockwell. Article "Mining," in Lalor's "Cyclopedia of Political Science."

LAW REPORTS of California, 1850–1880; also of Arizona, Oregon, Washington Territory, Idaho, Montana, Colorado, and Utah. Legislative enactments relating to mines in California, Arizona, Idaho, and Montana.

MINERALS AND MINING.—Government Reports on Mineral Resources of the United States for 1867, (Brown and Taylor), 1868 (same authors), and 1869 (Prof. Raymond). Hittell's History of "Mining in the Pacific States." Reports of Clarence King. Files of Engineering and Mining Journals, 38 vols., N. Y. Resources of California, (Hittell). Senate of California Reports, 1853–4. Bean's "Nevada Directory," 1867. "Natural Wealth of California," (Cronise). "Mines of the West," (Raymond).

"OVERLAND MONTHLY."—First Series and Second Series, San Francisco, nineteen volumes; also "*The Californian*," four volumes; also the "*Hesperian*," one volume; also "*Hutching's Magazine*," four volumes; also files of the *Mining and Scientific Press*, the *Alta Californian*, and the Daily *Bulletin*, all of San Francisco; also scraps from Pacific Coast local journals extending over a number of years.

PIONEERS, CORRESPONDENCE WITH.—Letters from the editors of the Boise City, Idaho, *Republican;* the Nevada City, California, *Transcript*, and other local journals, including interviews with old miners. Letters from Messrs. Dorsey (Grass Valley), Roberts (Shasta), Marshall (Trinity); Wilcox, Blake, Richardson, Dougherty, and others in San Francisco; Clough and Lynch of Alameda; also, from local county officers in Arizona, Idaho and Montana.

SPEECHES.—Of Senator Stewart and others in Congress over proposed Legislation, 1865–1866. Report of Senator Conners, 1866.

"*Vigilantes of Montana.*" Dimsdale. Helena, M. T., 1866.

INDEX TO SECOND VOLUME

OF

JOHNS HOPKINS UNIVERSITY STUDIES

IN

HISTORICAL AND POLITICAL SCIENCE.

C

D

E

F

G

M

N

O

P

Q

R

S

W

Y

Made in United States
Troutdale, OR
11/03/2026

56844449R00049